John, who makes me look again — 2016. tx.

# Sacred Ibis

# Sacred Ibis

*The Ornithology of
Canon Henry Baker Tristram, DD, FRS*

— W. G. HALE —

Sacristy
Press

**Sacristy Press**
PO Box 612, Durham, DH1 9HT

www.sacristy.co.uk

First published in 2016 by Sacristy Press, Durham

Copyright © W. G. Hale 2016
The moral rights of the author have been asserted

All rights reserved, no part of this publication may be reproduced or transmitted in any form or by any means, electronic, mechanical photocopying, documentary, film or in any other format without prior written permission of the publisher.

Scripture quotations are from Revised Standard Version of the Bible, copyright © 1946, 1952, and 1971 National Council of the Churches of Christ in the United States of America. Used by permission. All rights reserved.

All photographs were taken by the author except where they are credited to others.

Every reasonable effort has been made to trace the copyright holders of material reproduced in this book, but if any have been inadvertently overlooked the publisher would be glad to hear from them.

Sacristy Limited, registered in England & Wales, number 7565667

**British Library Cataloguing-in-Publication Data**
A catalogue record for the book is available from the British Library

ISBN 978-1-910519-13-4

**Frontispiece:** Dr H. B. Tristram (1862) as he was pictured in the Jubilee Supplement of the *Ibis*, 1908, Vol. 50:153, two years after his death. Reproduced with the permission of the British Ornithologists' Union (www.bou.ac.uk).

*In Memory of James Birkett Cragg,*
*Professor of Zoology, University of Durham, 1950–1961*

*"Canon Henry Baker Tristram, FRS—The most important*
*biological scientist to have emerged from Durham"*
*J. B. Cragg, 1992*

# Foreword

As one of a group of students surveying birds in Turkey we fantasised over ever seeing Tristram's Grackle, for which the field guide said, "Very vocal, has characteristic meandering whistles recalling tuning in short-wave radio, e.g. wioowiooweet, often in flocks around ravines", and I was delighted to see this remarkable bird decades later. I confess that beyond this, and being aware of a few other bird species bearing his moniker, I had no idea who he was.

I was thus impressed to learn about the scale of his travelling and collecting, that he was a pioneer in accepting the idea of evolution through appreciating how this explained many features shown by larks, such as their capacity to always be difficult to spot in any habitat, and furthermore that he was in the group, along with Alfred Newton (based in the department where I work), that established the British Ornithologists' Union.

I am delighted that Bill Hale has provided the opportunity to learn more about this remarkable man.

**William J. Sutherland**
*Miriam Rothschild Professor of Conservation Biology*
*Department of Zoology, University of Cambridge*

# Preface

Like many schoolboys during the Second World War, I collected birds' eggs and this was my introduction to ornithology. I gave my collection to Bolton Museum before the 1954 Bird Protection Act and before I went to University in Durham. I suppose that if I had lived in Victorian times I would have continued to collect birds' eggs, and probably skins too, for this is what Victorian ornithologists did. My introduction to Victorian ornithologists was through a copy of Henry Seebohm's *Coloured Figures of the Eggs of British Birds* which my father bought me for my seventh birthday and it was from this that I first identified my favourite bird, the Redshank. From books such as this and later Alfred Newton's *Ootheca Wolleyana*, came my interest in the history and development of scientific ornithology and antiquarian books on the subject. Later, first as a schoolboy and then as an undergraduate, I attended several of David Lack's post-Christmas Edward Grey Institute Conferences in Oxford, and talked long into the nights with great ornithologists such as Niko Tinbergen, Reg Moreau, Arthur Cain (later a friend) and David Lack himself. These conferences featured Alauda vespertina (the Evening Lark), which was an ornithological quiz. On more than one occasion a question featured clergymen interested in ornithology and there were many of them; in fact, it occurred to me that they were the professional ornithologists of Victorian times! One such was Canon Henry Baker Tristram, DD, FRS.

I was particularly interested in Alfred Newton, Professor of Zoology in the University of Cambridge, and I bought a bound collection of his papers, many of them signed by him as presentation copies to Tristram:

*Viro illustri primo reverendoque,*
*H. B. Tristram D.D. Auctor AN.*

On the front endpaper was Tristram's bookplate:

Sigill:Henrici:Baker:Tristram:e:Coll:Linc:Oxon.

In 1953, whilst still at school, I first discovered Liverpool Museum and became friendly with Reg Wagstaffe, Keeper of Zoology. During the Second World War, the drawers of the skin cabinets had been removed hurriedly during the bombing, which later destroyed many of the wooden frames. Even in 1954 everything was still in piles of drawers and it was impossible to locate particular skins. I helped from time to time to organize these and discovered collections of Hawaiian Honeycreepers, Darwin's Finches and other treasures, many of which formed part of the Tristram Collection acquired in 1896 by the Free Public Museums of Liverpool (now the World Museum, of Liverpool Museums Merseyside). Later, as a research student in Durham, I discovered, in the Zoology Department, a collection of some fifty or sixty skins bearing Tristram's labels. With the agreement of the University I later took these to join the main collection in Liverpool.

In 1956 Professor F. S. Bodenheimer, of the University of Jerusalem, visited Durham in order to find material for his book on Tristram which was published in Hebrew, in Jerusalem, and entitled *Canon Henry Baker Tristram (1822–1906): The Father of the Natural History of Palestine.* Professor J. B. Cragg, who was at the time Head of the Department of Zoology in Durham, gave Bodenheimer a great deal of help in finding material for the book. Bodenheimer (1957) described his visit to Durham as being "to absorb something of the atmosphere of Tristram's one-time home and surroundings". Of Tristram he also wrote that he was "one of the great traveller-naturalists of the nineteenth century. The results of his expeditions ... are of outstanding importance." He also listed Tristram's Obituaries and began a bibliography that was hindered by a lack of some of the relevant journals in the University library (Bodenheimer, 1956b).

Bodenheimer was an ecologist and lectured whilst he was in Durham. Many of us who went to listen to him remember him to this day.

Over the years I have looked for birds in many of Tristram's haunts. During the seven years I spent in Durham I visited Lindisfarne and the Farne Islands many times and Cheviot and Fenham Flats less frequently. I have walked in the valleys of all three main rivers of the North-East and explored the Tees marshes on many occasions, mainly Cowpen Marsh and Seal Sands, in search of wading birds. I have visited most of the coastline between St Abb's Head and the Flamborough Cliffs and photographed many of the birds of these places. Since leaving Durham I have regularly visited the Farne Islands, again sailing as we did in the 1950s, in the company of that great character Billy Shiels, now no more.

Abroad I have followed Tristram's footsteps, admittedly unknowingly, in Norway, Namibia, Switzerland, Italy, Portugal, Spain, Gibraltar, Tunisia, Algeria, Egypt, the Greek mainland and many islands, Turkey, China and Jordan. I have visited many of Tristram's islands such as Malta, Sicily, Madeira and the Canaries and many which he never visited but from which he obtained birds for his collection through exchanges. I have trapped many birds during research projects and for general ringing, all of which were subsequently released, but I cannot claim to have handled the twenty thousand or so which Tristram collected.

Before I started this project I knew something of the Canon, but as I have read his papers, his letters and his books I have developed a great liking for him. He was an incredible person, one of the great pioneers of ornithology; and whilst now we might frown on his collecting, nearly all of us involved in ornithology, as an aspect of zoology, would have done as he did and been collectors, had we been born in his time. Over the years I have come to admire Newton's work and his sense of humour, naming Tristram "Sacred Ibis" (Members of the British Ornithologists' Union are referred to as 'Ibises') and later, after his canonisation, "The Great Gun of Durham". Tristram was a great friend of Newton—count the extant letters between them—and was from the same mould. As his granddaughter wrote, he may not have been a great churchman but he was certainly a great ornithologist.

J. B. Cragg wrote of Tristram in a letter to and quoted by Baker (1992) that he was "the most important biological scientist to have emerged from Durham". It is with this Tristramic background that I have decided to set 'pen to paper' (but metaphorically, as I sit at my computer) to tell Tristram's story, not as a churchman, antiquarian or traveller, but as an ornithologist and one of the greats of the Victorian era and for that matter, any other era.

**W. G. Hale**
*December 2014*

# Acknowledgements

My particular thanks are due to John Collins for discussing and advising on different aspects of the text and also for finding some obscure references for me. My wife Marie also read through the text and provided helpful advice. My friends Christine Jackson, Hugh Hollinghurst, Tim Milsom and Jim Whitaker all provided me with much help and information and made material available to me from their libraries as did Ken Williams.

Descendants of Canon Tristram's family, Jocelyn Holland, Janet Elphinstone and Margaret Elphinstone, provided photographs and a great deal of information during the course of writing the book. James Riley made available to me Darwin's first letter to Tristram and gave permission to publish it and to place a copy in the Cambridge University library.

Most of Tristram's scientific papers appeared in the *Ibis*, the journal of the British Ornithologists' Union (BOU). Images from these papers, fully acknowledged in the captions, and quotations in the text, where they are fully referenced, are reproduced in this work by permission of the BOU, www.bou.org.uk. In this connection I would particularly like to thank Steve Dudley, Senior Administrator of the BOU, for his help and cooperation.

The Natural History Museum at Tring made relevant documents available to me and for this I have to thank Mrs Alison Harding. Dr Robert Prys-Jones and Mr Douglas Russell together with Ms Yolande Ferreira provided the excellent images of Tristram's Great Auk's egg. I am grateful to Glasgow Museum for the image of the Great Auk (Z.1978.77), particularly Mr Richard Sutcliffe and Ms Susan Pacitti. From the National Museum of Edinburgh Mr Mark Glancy and Ms Georgia Rogers made available Tristram-associated correspondence as did Mrs June Holmes at the Hancock Museum, Newcastle-upon-Tyne. The Tristram skin collection and associated papers in the Liverpool World Museum were made available by Mr Tony Parker and Dr Clem Fisher. The York Museums Trust gave me permission to photograph and reproduce the Moa and I am grateful to Mr Stuart Ogilvy for his help. The image of the Bermuda Petrel was taken by Dr Jeremy Madeiros and supplied by the Bermuda Conservation Services and here I thank Ms Mandy Shailer. The images of the Willow Warbler (Plate 13) and Kestrel (Plate 62) were provided by my friends David Hosking and Phil Collins respectively, and the image of the mummified Sacred Ibis by Ashley Cooke, courtesy of World Museum, National Museums Liverpool. My thanks are also due to the authorities of St John's College, University of Durham and to Mr Robert Gladstone for permission to reproduce Plate 15.

The Cambridge University Library gave access to the material in the Darwin Correspondence Project and to the Newton Correspondence, and Mrs Sheila Hingley and Mr Andrew Gray provided information from both the Cathedral and University libraries in Durham. Permission to reproduce the photograph of Professor J. B. Cragg was given by the National Portrait Gallery and the pages from the Select Committee Report on Wild Birds Protection 1873 from the authorities responsible for the Parliamentary Archives. I thank all these institutions and individuals for their help.

I would also like to thank Professor William J. Sutherland from Alfred Newton's old Department for writing the short Foreword; I doubt that either Tristram or Newton would have envisaged their efforts in initiating the conservation movement eventually resulting in there being a Professor of Conservation in the University of Cambridge.

Richard Hilton and Thomas Ball of Sacristy Press were immensely helpful throughout the production of the book and in maximising the financial contribution to the Grey College Trust, University of Durham, to which all royalties arising from this book are contributed. For so painstakingly editing my manuscript I should like to thank Stephen Greenhalgh, not only for his corrections but also for his numerous suggestions which improve the text.

**W. G. Hale**
*January 2015*

# Contents

Foreword . . . . . . . . . . . . . . . . . . . . . . . . . . . . . . . . . . . . . . . . . . . . . . . . . . . vi
Preface  . . . . . . . . . . . . . . . . . . . . . . . . . . . . . . . . . . . . . . . . . . . . . . . . . . vii
Acknowledgements . . . . . . . . . . . . . . . . . . . . . . . . . . . . . . . . . . . . . . . . xi
Lists of Plates, Figures and Tables . . . . . . . . . . . . . . . . . . . . . . . . . . . . xv
List of Abbreviations . . . . . . . . . . . . . . . . . . . . . . . . . . . . . . . . . . . . . . xx

1. In The Beginning. . . . . . . . . . . . . . . . . . . . . . . . . . . . . . . . . . . . . . . 1
2. A New World: Tristram in the Bermudas . . . . . . . . . . . . . . . . . . . . 9
3. Castle Eden to Canon: Birdwatcher to Ornithologist . . . . . . . . . 27
4. An Exaltation of Ornithologists . . . . . . . . . . . . . . . . . . . . . . . . . . 39
5. The Sands of the Sahara: the Western Desert . . . . . . . . . . . . . . . 51
6. The Sands of the Sahara: in Tunisia . . . . . . . . . . . . . . . . . . . . . . . 75
7. A Darwinian Conversion. . . . . . . . . . . . . . . . . . . . . . . . . . . . . . . . 83
8. The Long Winters of Palestine 1: South to the Gore (1858, 1863–4) . . . . . . . . . . . . . . . . . . . . . . . . . . . . . . . . . . . . . . . . . . . . . 105
9. The Long Winters of Palestine 2: the Retreat from Beersheba to Beirut. . . . . . . . . . . . . . . . . . . . . . . . . . . . . . . . . . . . . . . . . . . . . 123
10. Palestine: The Land of Moab (1872). . . . . . . . . . . . . . . . . . . . . . 143
11. Palestine Revisited: Egypt to Armenia (1881). . . . . . . . . . . . . . 153
12. Tristram's Islands. . . . . . . . . . . . . . . . . . . . . . . . . . . . . . . . . . . . . 165
13. Tristram in Japan. . . . . . . . . . . . . . . . . . . . . . . . . . . . . . . . . . . . . 201
14. Twilight of the Great Auk. . . . . . . . . . . . . . . . . . . . . . . . . . . . . . 209
15. "What's Hit is History, What's Missed is Mystery": The Collections of Skins . . . . . . . . . . . . . . . . . . . . . . . . . . . . . . . . . . . . . . . . . . . . 221
16. Ootheca Tristramiana: Collector to Conservationist. . . . . . . . . 243

17. Bird Protection and the Grassroots of Conservation......... 253
18. Last Letters............................................. 265
19. Tristram's Library ...................................... 269

**References** ................................................. 275
**Appendix 1: Tristram's Life Table** ........................... 284
**Appendix 2: Correspondence associated with Tristram** ........ 287
**Appendix 3: Tristram's Ornithological Publications** ............ 300

# Lists of Plates, Figures and Tables

## Plates

Frontispiece: Dr H. B. Tristram (1862), pictured in the Jubilee Supplement of the *Ibis*, 1908

1. Female Hen Harrier *Circus cyaneus*
2. The Golden Plover *Pluvialis apricaria*
3. Dotterel *Charadrius morinellus*
4. The Gannet *Morus bassanus*
5. The Fulmar Petrel *Fulmarus glacialis*
6. Audubon's Shearwater *Puffinus lherminieri*
7. Bermuda Petrel *Pterodroma cahow*
8. Green Heron *Buteroides striatus*
9. Hudsonian Curlew (Whimbrel) *Numenius phaeopus hudsonicus*
10. Red-winged Blackbird *Agelaius phoenicius*
11. Shag *Phalacrocorax aristotelis*
12. A male Eider Duck *Somateria mollissima* (St Cuthbert's Duck)
13. Willow Warbler *Phylloscopus trochilus*
14. Eurasian Jay *Garrulus glandarius*
15. Tristram as he dressed during his explorations of North Africa and Palestine
16. Sacred Ibis *Threskiornis aethiopicus*
17. Tristram during the late 1860s
18. Alfred Newton, first Professor of Zoology and Comparative Anatomy, Cambridge
19. Montagu's Harrier *Circus pygargus*

20. Glossy Ibis *Plegadis falcinellus*
21. Black-shouldered Kite *Elanus caeruleus*
22. Collared Pratincole *Glareola pratincola*
23. Lanner Falcon *Falco biarmicus*
24. Sakker Falcon *Falco cherrug*
25. Barbary Falcon *Falco peregrinoides*
26. Little Stint *Calidris minuta* (flock) and Curlew Sandpiper *Calidris ferruginea*
27. Tristram's Warbler *Sylvia deserticola*
28. Purple Heron *Ardea purpurea*
29. Red Kite *Milvus milvus*
30. Tawny eagle *Aquila repax*
31. Common Crested Lark *Galerida cristata arenicola*
32. Common Crested Lark *Galerida cristata macrorhynca*
33. Mourning Wheatear *Oenanthe lugens halophila* (*Saxicola halophila* of Tristram)
34. Red-rumped Wheatear *Oenanthe moesta* (*Saxicola philothamna* of Tristram)
35. Egyptian Goose *Alopochen aegyptiacus*
36. Pied Kingfisher *Ceryle rudis*
37. Osprey *Pandion haliaetus*
38. Goldfinch *Carduelis*
39. Palestine Sunbird *Nectarinea osea*
40. Tristram's Grackle *Onychognathus tristramii*
41. European Kingfisher *Alcedo atthis*
42. Nubian Nightjar *Caprimulgus nubicus tamaricis*
43. Stone Curlew *Burhinus oedicnemus*
44. Lapwing *Vanellus*
45. Shoveler *Anas clypeata*
46. Spur-winged Plover *Haplopterus spinosus*
47. Pallid Harrier *Circus macrourus*
48. White Stork *Ciconia ciconia*
49. Nuthatch *Sitta europea*, courtship feeding

50. Syrian Serin *Serinus syriacus* and Dead Sea Sparrow *Passer moabiticus*
51. Cyprus Warbler *Sylvia melanothorax*
52. Black-winged Stilt *Himantopus himantopus*
53. Skylark *Alauda arvensis*
54. Common Teal *Anas crecca*
55. Hoopoe *Upupa epops*
56. The Mastaba of Meidoum
57. Rose-coloured Starling *Sturnus roseus*
58. Common Redshank *Tringa totanus*
59. Kentish Plover *Charadrius alexandrinus*
60. Darter *Anhinga rufa*
61. Dodo *Raphus cucullatus*
62. Common Kestrel *Falco tinnunculus*
63. Vegetarian Finch *Platyspiza crassirostris*
64. Woodpecker Finch *Camarhynchus pallidus*
65. South Island Moa *Dinornis robusta*
66. Takahe *Porphyrio mantelli*
67. Tristram about the time of his visit to Japan
68. Great Auk *Pinguinus impennis*
69. Norfolk Island Parakeet (Kaka) *Nestor meridionalis productus*
70. Labrador Duck *Captorhynchus labradorius*
71. Large Ground Finch *Geospiza magnirostris*
72. Warbler Finch *Certhidia olivacea*, the smallest of Darwin's Finches
73. Duchess Lorikeet *Charmosyna margarethae*
74. Bare-legged Scops Owl *Otus insularis*
75. White-collared (Mangrove) Kingfisher *Todiramphus chloris tristrami*
76. Tristram's egg of the Great Auk *Pinguinus impennis*
77. Eggs of the Greenshank *Tringa nebularia*
78. Eggs of the Great Snipe *Gallinago media*
79. Eggs of the Bar-tailed Godwit *Limosa lapponica*
80. Plaque in Memory of John Wolley in Southwell Cathedral, Nottinghamshire

81. A page from Tristram's Egg Catalogue
82. Nesting Seabirds on the Farne Islands
83. Kittiwake *Rissa tridactyla*
84. Title page from the report of the Select Committee
85. The beginning of Tristram's evidence to the Select Committee
86. "Climmers" collecting eggs on the cliffs at Bempton
87. Sale Catalogue of Tristram's books
88. Mummified Sacred Ibises

# Figures

Fig. 1. Original Members of the British Ornithologists' Union .... 42
Fig. 2. Title Page of the first volume of the *Ibis*, 1859............. 45
Fig. 3. Newton's note to Tristram ............................. 91
Fig. 4. Darwin's first letter to Tristram ......................... 99
Fig. 5. Map of the northern part of Palestine ................. 141
Fig. 6. Map of the southern part of Palestine.................. 142
Fig. 7. Map of the Dead Sea area........................... 163

# Tables

Table 1. Breeding Birds of Bermuda ........................... 21
Table 2. Introduced Breeding Birds of Bermuda.................. 22
Table 3. Scientific Names of Birds of Bermuda and North America.. 23
Table 4. Bird skins collected by H. B. Tristram in Bermuda 1847-9.. 24
Table 5. Birds observed by Tristram in North Africa .............. 65
Table 6. Dimensions of Eggs of *Sitta*........................ 140
Table 7. Mascarene Islands: Endemic species of birds ............ 194

| | | |
|---|---|---|
| Table 8. | Seychelles Endemics | 195 |
| Table 9. | Galapagos: Endemic species of birds | 196 |
| Table 10. | Hawaiian Endemic Birds | 198 |
| Table 11. | Second Tristram Collection | 233 |
| Table 12. | Tristram's Birds | 238 |

# List of Abbreviations

| | |
|---|---|
| **BA** | British Association (for the Advancement of Science) |
| **BOA** | British Oological Association |
| **BOC** | British Ornithologists' Club |
| **BOU** | British Ornithologists' Union |
| **BP** | Before the Present |
| **DCP** | Darwin Correspondence Project |
| **FRS** | Fellow of the Royal Society |
| **NHM** | Natural History Museum (London & Tring) |
| **NNR** | National Nature Reserve |
| **PEF** | Palestine Exploration Fund |
| **RSPB** | Royal Society for the Protection of Birds |
| **RSPCA** | Royal Society for the Prevention of Cruelty to Animals |
| **SSSI** | Site of Special Scientific Interest |

— 1 —

# In The Beginning

KITE. (*Milvus regalis.*)

How nice it must have been to be born in Northumberland. Had Henry Baker Tristram chosen this for himself, he would probably have thought that he could have done no better. So it was that he saw the first light of day on 11 May 1822 as the eldest son of the Reverend Henry Baker Tristram, vicar of Eglingham, near Alnwick, and his wife Charlotte. Whellan (1855) describes Eglingham as exhibiting "a great variety of soil and scenery, from the sterile moor to the fertile, highly cultivated valley and possesses a mineral spring, tinctured with sulphuric acid and which issues from an old drift for the draining of coal pits; as also a lake covering nine acres, called Kimmer Lough, abounding in perch and pike. . . . there are some vestiges of British and Roman encampments and the ruins of an old border tower". Eglingham lies in the valley of the Eglingham Burn, a tributary of the River Aln, and some five miles from the Cheviot Hills—Elysium for a small boy interested in natural history.

As Henry grew older his interest in natural history was encouraged by friends of his father, particularly Ralph Carr (later Carr-Ellison) of Dunston Hill and J. C. Langlands of Old Bewick, who took him on field trips away from his home village. Coquet Island at the mouth of the Aln, the Farne Islands, Lindisfarne (Holy Island), the Cheviot Hills and Prestwick Carr are all sites that the young Tristram would have visited and like many other small boys, he collected birds' eggs. Much later in life (Tristram, 1889a), when he was preparing his *Catalogue of a Collection of Birds* (1889), he commented in the preface that the collection dated from 1844 and that "the few specimens of my schoolboy days . . . have disappeared". Bowdler Sharpe (1906) writing on the young Tristram's ability as a field naturalist, along lines which perhaps we would not now applaud, informs us that "before the age of fifteen he had taken with his own hands, and within a walk of his home, the eggs of the Kite, Buzzard, Marsh Harrier, Hen Harrier, Peregrine Falcon and Raven". This tells us also that he was something of a climber and contradicts the impression that records of his future ill health might give about his fitness. Bolam (1912) comments on several of these trophies: of the Kite eggs Tristram is recorded as having "himself taken nests in Eglingham Wood, in his younger days", i.e. before 1837; of the Common Buzzard, "Canon Tristram told me, in 1890, that in his young days he had taken nests in Eglingham and Beanley Woods and that a small cliff on the edge of the latter moor was a regular nesting site for many years, not finally deserted till near the middle of the nineteenth century"; of the Marsh Harrier, "the late Canon Tristram has frequently told me of nests which he had taken at Kimmer Lough where the bird nested regularly when he was a boy"; and of the Hen Harrier (Plate 1), "Canon Tristram has told me of nests he used to see, and take, in more than one place on the moors in that neighbourhood during his residence at Eglingham Vicarage." Of the Peregrine Falcon and the Raven, Bolam does not quote Tristram, as both species were reasonably common long after his depredations; however, he does write of the Peregrine: "Egg-stealers are also accountable for the destruction of occasional nests"! In no way was Tristram responsible for the decline of any of these species no matter how we might view his past activities now. He had a very happy boyhood and surrounded by some of the best countryside in England he had wandered far and wide on his pony and on foot over the moors and along the shores. He had taken an interest in all around him, particularly the birds, and he had developed a great interest for the outdoors.

Tristram's education had been the responsibility of his father and his father's curate, and this had benefited him greatly, resulting in his being significantly more advanced academically than most boys of his age. In 1830 Tristram's mother died after the birth of her sixth child. In all there were two sons and four daughters and according to Tristram's daughter Louisa "from this time little Henry had become his father's constant companion" (Tristram, L. H-H., 1898). He started a journal which he kept for only a few years and which he resurrected only on his expeditions, leaving no extensive record of his activities at other times.

In his journal Tristram recorded his enjoyment of the family dogs and his pony, and particularly a three-months-long excursion with his brother and father around England. From Durham, York and Grantham, where his father had been born, they travelled to London and spent some time there, then on to Weymouth, Bath and Bristol, through Shrewsbury and back to Eglingham. Tristram clearly enjoyed this excursion very much and it probably instilled in him the desire to undertake his future wanderings. About this time Tristram's father decided to marry again to Miss Ann Wood of Edinburgh. She proved to be an excellent step-mother and Louisa recorded that "once again it was a very happy home".

There are few places like Eglingham which are within easy reach of so many different habitats. Fields, forests, some of the best cultivated land in the country contrasting with wild stretches of moorland and mountain and, of course, the sea shore with its cliffs, islands, mud-flats and sandy beaches are all still there. In addition there are lakes, bogs and fenland, so that there can have been few places in Britain to appeal more to the young naturalist.

Some of these places have changed over the years since Tristram knew them as a boy. Kimmer Lough remains, but the Marsh Harrier is long gone. There is still an area of open water of approximately four hectares but the surrounding wilderness of the middle 1800s has been replaced by agricultural land. Prestwick Carr is largely drained, and the raised bogs no longer to be found, but a small area remains as a Wildlife Trust marshland. Plans exist to develop this within the 'Living Landscapes' initiative, to join other such areas but this is inhibited to some extent by part of the original carr now being Ministry of Defence land. Inland from Eglingham, Cheviot lost its Golden Eagles before Tristram's time but the falcons and the ravens remain amongst the crags and cliffs. On the moors Golden Plover (Plate 2) are not uncommon and a few breeding Dotterel (Plate 3) remain.

Elsewhere away from the coast all is not doom and gloom. Like the Marsh Harrier, the Hen Harrier no longer haunts the area in spring, but Goshawk and Honey Buzzard now breed in Northumberland and the programme for the reintroduction of the Kite is proving successful.

Whilst Tristram might have been disappointed by what has happened to some of the habitats that have disappeared since his time, he would be well pleased with the status of the birdlife, both inland and along the coastline. Along the north coast the Gannets (Plate 4) of Bass Rock have flourished and this has given rise to another increasing colony to the south, on Flamborough Head, in North Yorkshire. As a result Gannets are to be seen much more commonly off the coasts of Durham and Northumberland than in Tristram's time. The seabird colonies on the Isle of May, St Abbs Head, the Farne Islands, Dunstanborough and at Flamborough are flourishing, with only occasional setbacks over the years. On Lindisfarne (Holy Island) and Coquet Island little has changed except that now there are very few

Roseate Terns on the former but the Coquet Island population reached eighty pairs in 2010 (Holling et al., 2012).

Nowadays the mudflats remain much as they were a hundred and fifty years ago and passage and wintering flocks of waders and wildfowl are not seriously threatened. On the whole Tristram's birds have survived well on what are now carefully conserved habitats whose future is more secure than in the Canon's time. One thing that would have pleased him very much is the arrival on the British cliffs of the Fulmar Petrel as a breeding bird. Up to 1816, and possibly 1839, St Kilda was the only breeding place of this species in the north Atlantic south of Iceland (Holloway, 1996). It bred for the first time on the Farne Islands in 1935, and about the same time south of St Abbs Head. Probably Tristram saw Fulmars (Plate 5) flying off the coast, but it was twenty-nine years after his death that they first bred and his collection of skins contains only specimens taken from St Kilda and Iceland. Now there are nearly two hundred and fifty pairs breeding on the Farne Islands.

Easy access to these wild places was not to last, for in May of 1837 Tristram's father died and the family had to leave their old home in Eglingham. It was decided that they should make their new home in Durham and Mrs Tristram moved with the six step-children into a house in Crossgate which much later became the Parish Rooms.

When he was fifteen years old, and shortly after his father died, Tristram was sent to Durham School. He entered in 1837 and left in the December of 1839, aged seventeen. A few years previously, in 1833, the new University of Durham (the third university to be established in England) had decided to create a museum and this was housed on the banks of the Wear in the Old Fulling Mill. William Proctor was appointed its first Keeper at an annual salary of £25. Originally a carpenter's apprentice he was particularly interested in birds and was a competent taxidermist. It did not take long for Tristram to discover this and it was from Proctor that he first learned to make a bird skin and later mounted specimens. In Proctor's workshop above the museum, Tristram produced his first mounted bird, a Jackdaw (Mearns and Mearns, 1988), which he gave to Proctor.

Tristram had several bouts of illness when young but between these he seems to have been very fit. Louisa records that he always ran down Crossgate (never walked) on his way to school and on one occasion, when a horse was led across his path, he placed his hand on the animal's neck and vaulted over it—"the energy bred in him in the Northumberland moors never left him!". School was a new experience for him being in the company of so many boys when previously he had only a few companions in his country home.

On leaving Durham School, and apparently being too young to enter Oxford, Tristram spent two terms in the then new Durham University, proceeding to Lincoln College, Oxford in the autumn of 1840. The probability is that he spent most of this time in Durham with William Proctor, in the museum, honing his new skill of taxidermy. By now Tristram had come into contact with most of the ornithologists in the North-East of England and in a letter to Bowdler Sharpe in 1906, the last year of his life, he listed them as Jardine, Selby, Yarrell, Salmon, Johnstone and the Hancocks. Of this list Baker (1996) comments that "Tristram might have been referring to John Drew Salmon (1802–1859), ornithologist and botanist, who began to make an egg collection about 1828 and to George Johnston (1797–1855), spelt without an 'e', of Berwick-upon-Tweed, a prominent figure of influence in natural history in the North-East of England at the time Tristram was a young man".

In October 1839, Tristram matriculated in Oxford but did not enter the university until six months later. On his first visit to Oxford he carried a letter of introduction from Mr Faber, Master of Sherburn Hospital, to his nephew Mr Frederick Faber, a Fellow and tutor at University College. On their meeting Tristram was invited to dine with his new acquaintance and afterwards "taken to wine at Oriel with Mr Froude". Here he was introduced to a company consisting of Mr Sidney Herbert, Mr W. E. Gladstone and Mr Roundell Palmer (afterwards Lord Selborne). During the subsequent conversation Gladstone made derogatory remarks about Protestantism as a result of which Tristram retained a strong prejudice against him for the rest of his life, at the same time much admiring Disraeli whom at that time of the Oriel gathering he had not met (Tristram, L. H-H., 1898).

There is little of ornithological interest on record of Tristram's time at Oxford from where he graduated in 1844 with a degree in Classics. Since his printed Catalogue begins only in 1844 there is no listing of specimens collected before that date and thus no indication of where he might have spent his leisure time. Tristram spoke little of his time at Oxford where he made few close friends but he was fond of hunting and boating and excelled at both, his ambition being "to shine in the schools, on the river and in the field".

His room boasted a stuffed hedgehog in the centre of the mantelpiece, a falcon at one end and an owl at the other. Louisa records that "he had the best library any undergraduate at the time possessed, my father having inherited many of his father's books and being a great book fancier himself." After leaving Oxford he spent some time in Switzerland and in Italy as his health was not good, his illness being diagnosed variously as weak chest, bronchitis, ague and tuberculosis (Baker, 1996). Tristram arranged to tutor a Mr Charles Harford who was to tour with his parents on the continent, spending six months in Switzerland and a year in Italy. After making these arrangements Tristram was not able to join the family straight away as he had examinations to take. He subsequently travelled to Switzerland alone, spending three days in Paris and continuing the journey "by diligence through Dijon to Geneva".

It was in Geneva that Tristram had what might be considered the first of his many adventures. On the way to church one Sunday, Tristram and Charles Harford, the eldest son of the family and his tutee, espied a very collectable butterfly (species not recorded) and followed it, in an attempt to collect it, into an adjoining meadow—"unfortunately the strictly preserved property of Monsieur Pictet de Pregez". Here the intending collectors were themselves collected—arrested by two gendarmes who led them past the English gathering of potential Protestant worshipers, to be presented to the Mayor. The Mayor was "shaving and eating strawberries" at the time of the presentation, and continued to do so whilst he listened solemnly to the charge that they had been arrested "à la chasse de papillon". Very sensibly he said "Cinque francs d'amande" and set the prisoners free! (Tristram, L. H-H., 1898).

In Switzerland Tristram began his collecting and initially he was very selective. Away from Geneva, where again he was improving his taxidermy, only two specimens are recorded in his Catalogue; an Eagle Owl *Bubo bubo* was taken on the St Bernard Pass and an Alpine Accentor *Prunella collaris* on the Simplon Pass, both in 1844. About this time he must also have passed into France as there, in Chamonix, he records taking a Yellow-billed (Alpine) Chough *Pyrrhocorax graculus*. From the environs of Geneva his Catalogue contains records of only eight specimens, Greenshank *Tringa nebularia*, Little Stint *Calidris minuta*, Hoopoe *Upupa epops*, Rock Thrush *Monticola saxatilis*, Wall Creeper *Tichodroma muraria*, Nutcracker *Nucifraga caryocatactes*, Golden Oriole *Oriolus oriolus* and Ortolan Bunting *Emberiza hortulana*. Only the first two and the last of these could he have hoped to collect in England.

In Italy, in 1845, Tristram was apparently even less active in his collecting, taking only a Scops Owl *Otus scops* in Sienna (dated 1844 but probably mistakenly) and a Skylark *Alauda arvensis* in Turin. Tristram exchanged many skins with other ornithologists and it is possible that this happened to some he collected in Switzerland and Italy, though most of his exchanges would be of species more difficult to obtain, such as those he collected in the Middle East.

Later in 1845 Tristram was ordained Deacon by the Bishop of Exeter and appointed to a curacy of Morchard Bishop, some ten miles north-west of Exeter, in Devon. Appointed then as Priest he took up residence in Devon but over the next year he was again troubled by his chest condition. He continued to add to his skin collection but in his Catalogue there are only eight entries for this period of his life. He took Hen Harrier *Circus cyaneus* on Exmoor, Stonechat *Saxicola rubicola* around Torquay, and Grey Heron *Ardea cinerea*, Spotted Crake *Porzana porzana*, Great-spotted Woodpecker *Dendrocopus major*, Green Woodpecker *Picus viridis*, and Redstart *Phoenicurus phoenicurus* near Morchard Bishop. Tristram's illness persisted and he was advised to spend time in a warmer climate. Thus it was that towards the end of 1846 he accepted an appointment as secretary to Sir William Henry Elliot, the Governor of Bermuda. This post also involved his acting as naval and military chaplain, a duty that was to play a very significant part in Tristram's future as an ornithologist.

— 2 —

# A New World: Tristram in the Bermudas

ROCK PIGEON. (*Columba livia*.)

It was towards the end of 1846 that Tristram arrived in the Bermudas after what was a slow but relatively uneventful journey. The Bermudas, or Bermuda as the islands are now called, are a group of some 150 volcanic islands, topped by limestone formed from corals. These lie some 560 miles (900 km) east of South Carolina, and were firstly a Company Colony, then a British Crown Colony in Tristram's time, but are now a British Overseas territory. The mild humid climate is maintained by the Gulf Stream which flows between the volcanic islands—which have arisen from the sea bed—and the eastern border of the U.S.A. Alternatively known as the Somers' Islands at the time of Tristram's arrival, there was little known of their ornithology.

The Bermudas were discovered in 1505 by Juan de Bermudez but were uninhabited and not colonized until 1610, more by accident than design, and the story is of shipwreck. A flotilla of nine ships left Plymouth on 2 June 1609 with some 600 passengers heading for the new colony of Jamestown, Virginia, and the flagship 'Sea Venture' was captained by Admiral Sir George Somers, after whom the Bermudas were first named by the colonists. Living up to its name, the flagship was separated from the rest of the flotilla in a severe storm. The ship, the first to be built specifically as an emigration vessel, leaked badly, and having thrown its twenty-four guns overboard, Admiral Somers beached the ship between reefs, on the first land sighted, without loss of life. The Bermudas, or Somers' Islands, were thus occupied for the first time. Eventually most of the passengers and crew arrived in Virginia on 23 May 1610, in two small ships built from the remains of the 'Sea Venture', but inevitably had to spend some months on the islands where they had to forage for much of their food. Amongst those who had been on board the 'Sea Venture' were Sylvester Jourdain and voyages clerk William Strachey who was Secretary elect of the Virginia Company. The point in mentioning the discovery of the Bermudas is that these two 'sea venturers' were to be the first to document any of Bermuda's birds, and one in particular, the Cahow, that puzzled Tristram during his residence there.

In a letter which was eventually published in Purchas (1625), Strachey wrote:

> A kinde of webe-footed Fowle there is, of the bignesse of an English Greene Plover, or Sea Meawe, which all the summer we saw not, and in the darkest nights of November and December (for in the night they only feed) they would come forth, but not fly farre from home, and hovering in the ayre, and over the Sea, made a strange hoolows and harsh howling. They call it of the cry that it maketh, a

Cohow. Their colour is inclined to Russet, with white bellies, as are likewise the long feathers of their wings. Russet and White these gather themselves together and breed in those Lands which are high, and so far alone into the Sea, that the Wilde Hogges cannot swim over them, and there in the ground they have their Burrowes, like Conyes in a Waren, and so in the loose Mould, though not so deepe; which birds in a light bough in a darke night (as in their Lowbelling) we caught, I have been at the taking of three hundred in an hour, and we have laden our Boates. Our men found a prettie way to take them, which was by standing on the Rockes or Sands by the Sea-side, and hollowing, laughing, and making the strangest outcry that possibly they could; with the noyse whereof the birds would come flocking to that place, and settle upon the very armes and head of him that so cryed, and still creepe neerer and neerer, answering the noise themselves; by which our men would weigh them with their hand, and which weighed heavyest they took for the best and let the others alone, and so our men would take twentie dozen in two hours of the chiefest of them, and they were good and well relished Fowle, fat and full as a Partridge. In January we had great store of their Egges, which are as great as an Hennes Egge. There are thousands of these Birds, and two or three Lands full of their Burrowes whether at any time (in two houres warning) we could send our Cockboat, and bring home as many as would serve the whole Company: which birds for their blindness (for they see weakly during the day) and their crying and whooting, wee called the Sea Owle; they will bite cruelly with their crooked bills.

Jourdain (1610) gave a similar account, almost certainly of the same species of bird:

Another Sea fowle there is that lyeth in little holes in the ground, like unto Coney holes, and are in great numbers exceedingly good meate, very fat and sweet (those we had in the winter) and their eggs are white, and of that bignesse, that they are not to be knowne from [hens?] egges. The other birds' eggs are speckled and of a different colour.

The Rev. Lewis Hughes, later writing from the 'Somers Islands' in December 1614 comments on the same bird:

> Here is also plenty of sea foules, at one time of the year, as about the middle of October, birds which are called Cahouze and Pimlicoes come in. The Cahouze continue till the beginning of June in great abundance, they are bigger bodied than a Pigeon & of a very firm & good flesh. They are taken with ease if one do but sit down in a darke night, and make a noise, there will more come to him than he shall be able to kill: some have told me that the (islanders?) have taken twelve or fourteen dozen in an hour. When the Cahouze time is out, other birds called noddies and sandie birds come in, and continue till the latter end of August (in A. E. Verrill, 1901).

It would appear from these early comments that there were two species of 'night birds', at least one of which laid large white eggs and nested in burrows in the ground, and 'day birds', noddy and other terns (sand birds) nesting in colonies and laying speckled eggs. Further information is derived from John Smith's *Generall Historie of Virginia, New England and the Somer Isles* (1624). According to Jardine (1848–53) Smith stated that many small birds were killed by 'Wilde cats'—presumably feral cats—and he also comments on the Cahow. Documents in the collection of Sir Hans Sloane in the British Library (Sloane MS 750) seem to indicate that these comments were made by Nathaniel Butler, Governor of the archipelago between 1619 and 1622. Smith never visited the Bermudas but the documentation is still referred to in the literature under his name (Smith, 1624). A summary account of the history of the Cahow was given by Verrill (1901) but his conclusion that the bird was an auk was erroneous. Tristram, in his time in Bermuda had had other ideas nearer to the truth. Now the Cahow is known to be the Bermuda Petrel *Pterodroma cahow*.

Sometime after Tristram's arrival on the islands two contingents of the 42nd Highland Regiment (later the Black Watch) were drafted to the colony in the spring of 1847, amongst them two officers interested in ornithology. It is possible that Tristram already knew Henry Maurice Drummond before he arrived in Bermuda and possibly he knew him personally. Drummond (the future Colonel Henry Maurice Drummond-Hey who was to become the first Chairman of the British Ornithologists' Union) was a very competent field naturalist who took a great interest in the birds wherever he happened to be stationed. The second ornithologist was Lieutenant Wedderburn who apparently spent most, if not all, his spare time studying the birds of Bermuda, and it was he, together with Tristram, who provided information to Sir William Jardine for an article—in Jardine's own short-lived and obscure journal, *Contributions to Ornithology* 1848–53—on the Birds of Bermuda (Jardine, 1849). Throughout his time in the Bermudas Tristram carried out a continuous correspondence with Jardine, though he never published his results under his own name. In fact at this time Tristram had no ornithological publications to his credit.

When earlier travelling in Europe Tristram would have known most of the birds that he saw, having been familiar with them in England. In the Bermudas he had no such advantage as he encountered an avifauna largely related to the New World and he was not to be aided by his military contacts until some months later. However, though there is no record of this, it is unlikely that Tristram would have made the journey across the Atlantic without being prepared. His contacts in England would have advised him what to expect and he would certainly have been aware of the various editions of Alexander Wilson's *American Ornithology* (1808 to 1853) and possibly Audubon's *Ornithological Biography* (1831–9). Making identification easier for him was the fact that he was probably not expecting to encounter too many resident species; in practice he was surprised to find so few. Jardine's paper records 111 species for which Tristram and Wedderburn provided information. Of these, five were considered to be of doubtful identity and only eleven proved to be breeding species. In Tristram's Catalogue of his collection, seventeen other species are recorded, details of which had not been conveyed to Jardine, giving a grand total of 128 species recorded in the Bermudas by Tristram and Wedderburn. Over the years 375 species have now been recorded on the islands, only thirteen of which are considered to be birds breeding there at the present time (Table 1). Eleven more were accidental human introductions (Table 2) and of these the Yellow-crowned Night Heron *Nyctinassa violacea* was deliberately re-introduced allegedly in order to control land crabs.

Considering first the naturally breeding birds, the one of most interest is the Cahow. The most recent authority (del Hoyo et al., 1992) states: "Unknown for almost 300 years until rediscovered at Bermuda, where breeding confirmed 1951, with 18 pairs." In writing to Jardine, Tristram remarks in his printed list "that a few are still known to breed on or near Cooper's Island" so that if this is true the species was not unknown for 300 years!

Smith (1624) confused his account of the Cahow with his description of the nesting terns but it is clear that at this time the Cahow was very common, probably numbering tens (if not hundreds) of thousands (Brinkley, 2012). During difficult times of famine and fever in 1614 many of the colonists moved to Cooper's Island specifically to feed on these birds. Also elsewhere on the Bermudas the colonists took to eating both the birds and their eggs, so drastically reducing the population (Newton, 1896). However, there was an additional reason why the size of the breeding population of the Cahows was reduced and this was due to the large numbers of hogs that had been released previously by passing Spanish sailors. Initially the colonists fed on these hogs and reduced their numbers considerably, so, in one way, helping the birds. Even so, it is recorded in Smith (1624) that "Mr Norwood hath taken twenty dozon of them [the Cahows] in three or four hours, and since there had been such havocke made of them, they were near all destroyed, till there was a strict inhibition for their preservation." It would appear that this legislation had some effect as Purchas (1625) and Jones (1859) record their presence in November and December "coming forth only on the darkest nights" and being attracted to alight on the observers by "whooping and hallooing" in the manner described by Smith. If Tristram's comment to Jardine is correct, and there is no reason to believe otherwise, the Cahows were still present in Tristram's time.

In 1847, J. M. Jones visited several islands around Castle Harbour and was pleased "to learn from persons there resident, that the Cahow was still known by its old name . . . and that it still continued to breed in that locality." Specimens collected by Mr Salton Smith, of St George's in August 1847, were lost in the sea but were probably not Cahows as there were two fully grown chicks taken from the same nesting site. The Cahow was known, even then, to be unique in that it began its breeding season in late November and early December and normally laid only one egg. In May 1849, Captains Orde and McLeod climbed Black (Garnet Head) Rock close to Castle Harbour, and captured two fine specimens of the Dusky Shearwater, *Puffinus obscurus* of Audubon, now known as Audubon's Shearwater *Puffinus lherminieri* (Plate 6); one bird was sitting on a single white egg. Jones then came to a quite incredible conclusion, that the Cahow was, in fact, Audubon's Shearwater. This was a bird known to the Bermudan fishermen as the 'Pimlico', named by them, like the Cahow, because of its characteristic and distinct call. It was not the Cahow as claimed by Jones.

One of the two birds collected on 17 May 1849 by Captains Orde and McLeod was acquired by Tristram and is listed in his Catalogue of 1889, together with a reference to "Jones, Naturalist in Bermuda, p96". Tristram (1902a, 1902b) writes that he was in the company of Captain Orde when the birds and two eggs were collected fifty-three years previously (a fact not mentioned by Jones) when, in his opinion the problem of the Cahow was resolved.

Further interesting records were produced by S. G. Reid in two papers published in the *Zoologist* for October and November 1877, and later reproduced as a book (Reid, 1883). During his time serving with the army in the Bermudas he met Mr J. T. Bartram, a resident of the islands, who had made a collection of birds over a period of some twenty-five years. Bartram explained that on Cooper's Island there were "many nests", the Pimlico on the north side of the island and the Cahow on the south. He also confirmed the stories of the Cahow alighting on their watchers when the latter called out on dark nights, as had been recorded previously. Reid also recorded that he found two nests (of what?) in 1874 and tried to raise a chick which died after six weeks. He also recorded that the birds concerned alighted on him when he lay on the ground near the nest and that a single bird was shot by Private Hopegood in 1883, in daylight, crossing the harbour. From the behaviour of the birds and the fact that Reid "lay on the ground", it seems likely that he had found Cahows, whereas Hopegood had shot an Audubon's Shearwater.

No account seems to have been taken by any of the observers, that the Cahow and Pimlico were named separately by the local fishermen on account of their different call notes. There is no doubt that the birds collected by Orde, Tristram and Hopegood were Dusky Petrels, that is Audubon's Shearwaters. So were they Cahows or Pimlicos or are these different names for one species of bird? The answer was to be found four years after Tristram's death in 1906, when Louis L. Mowbray, Director of the Bermuda Aquarium, collected a live bird on Gurnet Rock, Cooper's Island. According to Brinkley (2012), who has documented subsequent events in detail, this bird was identified by Thomas Bradlee as the Mottled Petrel *Pterodroma inspectata* and there the matter lay for some ten years. In 1916 the skeleton of this bird was compared with skeletal remains found in the Crystal Cave in Bermuda and R. W. Schufeldt found that all the skeletal material was from the same species. Louis L. Mowbray and J. T. Nichols renamed the species *Aestrelata cahow*.

In 1935 William Beebe, of the New York Zoological Society, visited the Bermudas with the specific objective of finding other living birds of this species. In this he was not initially successful but was given a single bird that had flown into the light of the David's lighthouse. On sending this to the American Museum of Natural History in New York, R. C. Murphy identified it as a newly-fledged Bermuda Petrel.

Army officers in Bermuda seem to be inextricably bound up in the history of the Cahow as, in 1941, a bird that had flown into wires was sent by a member of the Volunteer Rifle Corps to Mowbray. The bird subsequently recovered and then was released. Again, in 1945, another army officer, Fred T. Hall, found the remains of several petrels at St David's, which were identified at the Smithsonian Institute, in Washington D.C., as Bermuda Petrels.

A search to establish the presence of nesting Cahows was carried out by Louis S. Mowbray (the son of Louis L. Mowbray), R. C. Murphy and his wife Grace Murphy in 1951, when seven pairs were found nesting around the islets in Castle Harbour. The first of these was found on 28 January 1951, in the presence of David B. Wingate, a young Bermudan ornithologist, who later led the conservation operation to preserve the colony of breeding Cahows. In between times he purchased a skin of the species from an antique dealer, and ironically, this had come from the collection of Mr J. T. Bartram, made during the early 1800s. Thus, Tristram's bird (Tristram, 1902a), which he died thinking was a Cahow, was in fact a Pimlico—Audubon's Shearwater *Puffinus lherminieri*—but he had almost certainly seen a Cahow in Bartram's collection!

It seems very likely that the Bermuda Petrel *Pterodroma cahow* (Plate 7), the Cahow of the fishermen, was never entirely absent from the Bermudas over the period of three hundred years or so during which it was supposedly unrecorded. The fact that the species has not been recorded breeding elsewhere supports this contention. Apparently the species hung on, probably in relatively small numbers, throughout this period because in winter it was permanently pelagic and during the breeding season ashore it was mainly, if not entirely nocturnal. This is in marked contrast with Audubon's Shearwater which is diurnal and normally nests in cracks in the rocks and not in burrows. Nowadays, this latter species is thought not to breed in the Bermudas though it is still listed on the check list of birds of the Bermudas by Avibase; it is a common species elsewhere.

With the facilities we now have to help with the identification of birds it is almost impossible to appreciate just how difficult it was in the mid-1800s going to a completely new country and being faced by a totally unfamiliar avifauna. It is clear that neither the Cahow nor the Pimlico were common or easy to find in Tristram's time in Bermuda, but he certainly played a part in what is a classic story of conservation. Now the calls of the Bermuda Petrel and Audubon's Shearwater can easily be compared on the internet; they are very different and it is not surprising that the Bermuda fishermen distinguished between the two species on the basis of their calls.

One of the most striking birds that Tristram found breeding in the Bermudas was what he named the Yellow-billed Tropicbird *Phaethon flavirostris* which is now known as the White-tailed Tropicbird *Phaethon lepturus*. Tropicbirds in general tend to nest in cracks in rocks, where there is plenty of room for their tails, but some choose to lay in burrows in the ground. This happened in the Bermudas where it caused some concern by competing with the petrels for their burrows. This was cleverly circumvented by Wingate providing artificial burrows which the Tropicbirds could not enter because of their greater size (Brinkley, 2012), and by providing artificial nesting sites away from the petrel nesting areas. Tristram (in Jardine, 1849) recorded that the Tropicbirds first arrived back at the islands around 9 March and left at the end of September (average dates).

The Green Heron (Plate 8), now recorded as a breeding bird in Bermuda, was not mentioned in Jardine (1849) but is recorded in Tristram's Catalogue as being shot by him in 1848 when he obtained two birds. Jones (1859) records Mr Hurdis obtaining one in October 1848 and the species being common in the spring and autumn of 1849. Like many other species it occurred in fluctuating numbers and in some years was present during the breeding season. It was only in 2002 that the first nests were recorded.

The Northern Bob-white *Colinus virginianus*—the Common Bob-white *Ortyx virginianus* of Tristram—was "formerly abundant when barley was more cultivated, but now nearly extinct and probably a bird of passage." (Jardine, 1849) Jones (1859) records it as breeding "thirty years ago" i.e., 1829, and also that Mr Richard Darrell tried to revive its fortunes in Bermuda by importing several pairs from the United States "to turn out". They were by the last accounts increasing rapidly. The species is not included in the present Avibase for Bermuda. The bird recognized by Tristram as the Florida Gallinule *Gallinula galliata,* now the Moorhen *Gallinula chloropus,* "is given in Mr Tristram's list amongst the constant residents" (Jardine, 1849) and is still a common breeding bird in Bermuda.

Terns were recorded in Bermuda in the publication attributed to Smith (1624) when they appeared to be so numerous and tame that both birds and eggs were taken in large numbers. It was not clear what species these birds were but from the observations of both Tristram and Wedderburn, Common Tern and Roseate Tern were present as breeding birds in the summer in the late 1840s. The former bred on the beach on Cooper's Island whilst the latter were to be found on "the farthest rock off St George's". Writing in 1883, Reid commented that whilst they had been common in

1850 no birds were present in 1874 and 1875. Nowadays the Roseate Tern is regarded as an Accidental Visitor whilst the Common Tern still breeds on the islands. The Roseate Tern last bred in the 1840s and the last attempt by the Least Tern was in the 1970s, (Dobson and Madeiros, 2006).

The Common Ground Dove and three species of Passerines make up the list of naturally breeding birds that Tristram found during his stay in the Bermudas. Whilst the dove seems to have held its own, all four of these species were adversely affected by the Cedar scale insect attack of the late 1940s. Between 1946 and 1953, 95 per cent of the 'Cedars' (not a true cedar, but a Juniper) were destroyed by the Juniper scale insects *Carulaspis minima* and *Lepidosaphes newsteadi*. The Eastern Bluebird was the most affected species since at that time it was nesting almost exclusively in holes in the cedars and had abandoned its previous habit of using holes in the cliffs. Competition for nest holes from introduced House Sparrows and European Starlings has also seriously affected the Bluebird. The White-eyed Vireo and Grey Catbird build cup-shaped nests in the few remaining trees and other vegetation but are preyed upon by the introduced Starlings and Great Kiskadee.

Tristram recorded a further four species as breeding birds but these have subsequently been found to have been introduced. The Mallard, the Rock Dove (Feral Pigeon) and the American Crow have all become common breeders whilst the recently reintroduced Yellow-crowned Night Heron became a regular breeder in 1980 (Wingate, 1982); the same author also records that the introduction had the desired effect as far as removal of land crabs was concerned in that the herons' diet was found to contain some 95 per cent land crabs.

Occasionally birds which are regarded as very infrequent visitors turn up in considerable numbers. In the autumn of 1814 "vast numbers" of Ruby-throated Hummingbirds descended on the Bermudas and "many remained until spring, particularly on David's Island, when they all disappeared, and have not been seen since." (Tristram in Jardine, 1849) The Cedar Waxwing appeared in flocks in 1847 and 1848 when they had not been previously recorded in the Bermudas, and the Snow Bunting was also present in flocks in the same winters (Jardine, 1849). Similarly, the Common Bob-White occasionally was recorded in considerable numbers as in January and February of 1847 and 1848, though it is commoner in the autumn (Jardine, 1849). When groups such as these occur on isolated islands there is a possibility of their remaining to breed and establish new populations. Tristram must have speculated along these lines. His later interest in the birds of island groups, particularly in oceanic islands, probably began when he lived in the Bermudas, and was stimulated by such irregular occurrences.

There can be little doubt that many of the winter visitors to the Bermudas are true migrants despite the fact that the islands lie so far from mainland America. Birds such as the Barn Swallow, Belted Kingfisher, several species of Herons, Waders and Warblers occur in most years, and whilst some may be driven to the islands by adverse winds and storms they keep returning. However, there are many species which are mere stragglers, forced from their normal migration routes by adverse weather, or just lost. Many of the Raptors and Owls fall into this category and the big majority are arrivals from North America. One or two are worthy of specific mention: the Eskimo Curlew is now extinct, but whereas Wedderburn (in Jardine, 1849) describes it as very rare, Reid (1883) considered it commoner than the Hudsonian Curlew (Plate 9): "A good number accompanied the Golden Plover on their arrival in September 1874, and several more were killed along the north shore." Tristram collected single individuals of each species in 1848.

Arrivals in the Bermudas from the east are rare. Jardine (1849) records two Common Snipe shot by Wedderburn in December 1847. A single Corncrake which flew into the lighthouse light on Somerset Island on 25 October 1847, which is in the Tristram Collection, is the only record of that species for Bermuda in Tristram's time there, as is the case with a single Wheatear (Jardine, 1849).

In May 1849 Tristram left Bermuda to return to his beloved North-East of England and to take up the Rectorship of Castle Eden, in County Durham. By this time his health was much improved and he decided to return home via North America. Whether this decision was determined by a wish to see more of the American birds of which he had had a taste in Bermuda or whether he saw it as an opportunity to see Niagara Falls we do not know. Certainly by this time Tristram was developing a taste for travelling, so his decision to see a little of North America was probably motivated both by the birds and a brief look at the continent of the New World. So it was that May 1849 found Tristram in the area of Niagara Falls. His gun travelled with his personal luggage and the first skin he added to his collection was that of a Great Northern Diver. In Canada, in the region of Niagara Falls, he was clearly selective in what he shot. Probably both opportunity and time to prepare skins restricted his activities somewhat, and in the next few weeks he collected only another eight species. Of the larger birds Tristram collected the Prairie Chicken and the Wood Duck. Northern Flicker, Blue-headed Vireo, Cedar Waxwing, Summer Tanager fell to his gun as did two species of crossbill, Red Crossbill and White-winged Crossbill. Crossing into the United States, to Buffalo, for a short time Tristram collected only two common species, Red-winged Blackbird (Plate 10) and Eastern Meadowlark.

As he sailed back across the Atlantic, in considerably better health than he had crossed in the opposite direction, Tristram was not to know that in the next few years he was to meet and befriend some of the greatest minds of his time and become involved in one of the biggest scientific controversies that his church and religion had ever faced. He was to become a country parson, and much more.

Table 1: Breeding Birds of Bermuda

| Present-day breeders | | Tristram's breeders* and his nomenclature |
|---|---|---|
| Pied-billed Grebe | *Podilymbus podiceps* | |
| Bermuda Petrel | *Pterodroma cahow* | *Pterodroma cahow*\* |
| White-tailed Tropicbird | *Phaethon lepturus catesbyi* | *Phaethon flavirostris*\* |
| Green Heron | *Buteroides striatus* | *Buteroides virescens* |
| | | *Ortyx virginianus*\* (N.Bobwhite) |
| Moorhen | *Gallinula chloropus* | *Gallinula galleata*\* |
| American Coot | *Fulica americana* | |
| Common Tern | *Sterna hirundo* | *Sterna hirundo*\* |

| Present-day breeders | | Tristram's breeders* and his nomenclature |
|---|---|---|
| | | Sterna dougallii* (Roseate Tern) |
| Common Ground Dove | Columbina passerina | Chamaepelia passerina* |
| Mourning Dove | Zenaida macrura | |
| Barn Owl | Tyto alba | |
| Grey Catbird | Dumetella carolinensis | Mimus carolinensis* |
| Eastern Bluebird | Sialis sialis | Sialia wilsonii* |
| Bermudan White-eyed Vireo | Vireo griseus bermudianus | Vireo noveboracensis* |
| 13 Species | | 10 Species |

Table 2: Introduced Breeding Birds of Bermuda

| Present-day breeders | | Tristram's breeders and his nomenclature |
|---|---|---|
| Yellow-crowned Night Heron | Nyctinassa violacea | Nycticorax violaceus |
| Mallard | Anas platyrhynchos | Anas boschas |
| Rock Dove | Columba livia | Columba livia |
| Great Kiskadee | Pitangus sulphuratus | |
| American Crow | Corvus brachyrhynchos | Corvus americanus |
| European Starling | Sturnus vulgaris | |
| Northern Cardinal | Cardinalis cardinalis | Guiraca cardinalis |
| European Goldfinch | Carduelis carduelis | |
| Orange-cheeked Waxbill | Estrilda melpoda | |
| Common Waxbill | Estrilda astrild | |
| House Sparrow | Passer domesticus | |
| 11 Species | | 5 Species |

Table 3: Scientific Names of Birds of Bermuda and North America

| Common Name | Scientific Name |
| --- | --- |
| Great Northern Diver | *Gavia immer* |
| Bermuda Petrel | *Pterodroma cahow* |
| Mottled Petrel | *Pterodroma inspectata* |
| Dusky Shearwater | *Puffinus obscurus* |
| Audubon's Shearwater | *Puffinus lherminieri* |
| Yellow-billed Tropicbird = White-tailed Tropicbird | *Phaethon flavirostris* *P. lepturus* |
| Green Heron | *Buteroides striatus* |
| Yellow-crowned Night Heron | *Nyctinassa violacea* |
| Wood Duck | *Aix sponsa* |
| Mallard | *Anas platyrhynchos* |
| Northern Bob-white = Common Bob-white | *Colinus virginianus* *Ortyx virginianus* |
| Prairie Chicken | *Tympanuchus cupido* |
| Florida Gallinule = Moorhen | *Gallinula galliata* *Gallinula chloropus* |
| Golden Plover | *Pluvialia dominica* |
| Eskimo Curlew | *Numenius borealis* |
| Hudsonian Curlew | *Numenius hudsonicus* |
| Common Snipe | *Gallinago gallinago* |
| Common Tern | *Sterna hirundo* |
| Roseate Tern | *Sterna dougallii* |
| Rock Dove (Feral Pigeon) | *Columba livia* |
| Common Ground Dove | *Columba passerina* |
| Northern Flicker | *Colaptes auratus* |
| Great Kiskadee | *Pitangus sulphuratus* |
| Cedar Waxwing | *Bombycilla cedorum* |
| Grey Catbird | *Dumetella carolinensis* |
| Eastern Bluebird | *Sialis sialis* |
| Wheatear | *Oenanthe oenanthe* |
| American Crow | *Corvus brachyrhynchos* |
| European Starling | *Sturnus vulgaris* |
| Snow Bunting | *Plectrophenax nivalis* |

| Common Name | Scientific Name |
|---|---|
| Summer Tanager | *Piranga rubra* |
| White-eyed Vireo | *Vireo griseus* |
| Blue-headed Vireo | *Vireo solitarius* |
| Red Crossbill | *Loxia curvirostra* |
| White-winged Crossbill | *Loxia leucoptera* |
| House Sparrow | *Passer domesticus* |
| Red-winged Blackbird | *Agelaius phoeniceus* |
| Eastern Meadowlark | *Sturnella magna* |

Table 4: Bird skins collected by H. B. Tristram in Bermuda, 1847–49.

| Current Nomenclature | Year | Qty | Tristram's Nomenclature |
|---|---|---|---|
| **Herring Gull** *Larus argentatus* | 1847 | 1 | |
| **Roseate Tern** *Sterna dougali* | 1847 | 1 | *Sterna* **dougalii** |
| **Grey Phalarope** *Phalaropus fulicarius* | 1848 | 1 | *Phalaropus* **fulicaria** |
| **White-rumped Sandpiper** *Tringa fuscicollis* | 1848 | 1 | ***Calidris*** *fuscicollis* |
| **Least Sandpiper** *Tringa minutilla* | 1848 | 2 | ***Calidris*** *minutilla* |
| **Buff-breasted Sandpiper** *Tryngites rufescens* | 1848 | 1 | *Trygites* **subruficollis** |
| **Spotted Sandpiper** *Trigoides macularius* | 1848 | 3 | *Actitis macularia* |
| **Greater Yellowlegs** *Totanus melanoleucus* | 1848 | 1 | *Tringa melanoleuca* |
| **Eskimo Curlew** *Numenius borealis* | 1848 | 1 | |
| **Ruddy Turnstone** *Strepsilas interpres* | 1848 | 3 | ***Arenaria*** *interpres* |
| **Semipalmated Plover** *Aegialitis semipalmata* | 1848 | 1 | *Charadrius semipalmatus* |

# A NEW WORLD: TRISTRAM IN THE BERMUDAS

| Current Nomenclature | Year | Qty | Tristram's Nomenclature |
|---|---|---|---|
| **Kildeer Plover** *Aegialitis vocifera* | 1848 | 1 | ***Charadrius*** *vociferus* |
| **American Golden Plover** *Charadrius virginicus* | 1848 | 4 | ***Pluvialis dominica*** |
| **American Coot** *Fulica americana* | 1848 | 1 | |
| **American Purple Gallinule** *Porphyrio martinicus* | 1848 | 1 | ***Porphyrula*** *martinicus* |
| **Common Moorhen** *Gallinula galeata* | 1848 | 2 | *Gallinula chloropus* |
| **Sora Rail** *Potzana carolina* | 1848 | 3 | |
| **Common Ground Dove** *Chamaepelia passerina* | 1849 | 20 | ***Columbina*** *passerina* |
| **Gadwall** *Anas strepera* | 1847 | 1 | |
| **Northern pintail** *Dafila acuta* | 1847 | 1 | ***Anas*** *acuta* |
| **Little Blue Heron** *Ardea caerulea* | 1848 | 1 | ***Egretta caerulea*** |
| **Great Blue Heron** *Ardea herodias* | 1848 | 1 | |
| **Green Heron** *Buteroides virescens* | 1848 | 2 | |
| **American Bittern** *Botaurus lentiginosus* | 1848 | 1 | |
| **Brown Pelican** *Pelecanus fuscus* | 1847 | 1 | *Pelecanus **occidentalis*** |
| **Yellow-billed Tropicbird** *Phaeton flavirostris* | 1848 | 3 | *Phaethon **lepturus*** |
| **Osprey** *Pandion haliaetus* | 1847 | 1 | |
| **Short-eared Owl** *Asio cassini* | 1847 | 1 | *Asio **flammeus*** |
| **Northern Saw-whet Owl** *Nyctala acadica* | 1849 | 1 | ***Aegolius acadicus*** |

| Current Nomenclature | Year | Qty | Tristram's Nomenclature |
|---|---|---|---|
| Yellow-billed Cuckoo *Coccyzus americanus* | 1848 | 1 | |
| Belted Kingfisher *Ceryle alcyon* | 1848 | 2 | ***Megaceryle*** *alcyon* |
| Yellow-billed Sapsucker *Sphyropicus varius* | 1848 | 2 | ***Sphyrapicus*** *varius* |
| Common Nighthawk *Chordeiles virginianus* | 1846–7 | 3 | *Chordeiles* ***minor*** |
| Eastern Kingbird *Tyrannus pipiri* | 1848 | 1 | *Tyrannus* ***tyrannus*** |
| Grey Catbird *Galeoscoptes carolinensis* | 1847 | 3 | ***Dumetella*** *carolinensis* |
| Eastern Bluebird *Sialia sialis* | 1846–9 | 7 | |
| Northern Waterthrush *Seiurus noveboracensis* | 1848 | 3 | |
| Nothern Parula *Parula americana* | 1849 | 1 | |
| Magnolia Warbler *Dendroeca maculosa* | 1848 | 1 | *Dendroica* ***magnolia*** |
| Blackpoll Warbler *Dendroeca striata* | 1848 | 1 | *Dendroica* *striata* |
| White-eyed Vireo *Vireo noveboracensis* | 1848 | 4 | *Vireo* ***griseus*** |
| Barn Swallow *Hirundo erythrogastra* | 1847 | 3 | *Hirundo* ***rustica*** |
| Northern Cardinal *Cardinalis virginianus* | 1848 | 6 | *Cardinalis* ***cardinalis*** |
| Bobalink *Dolichonyx orizivorus* | 1848 | 1 | |

Bold type in Tristram's nomenclature indicates differences to the current nomenclature.

Plate 1: Female Hen Harrier *Circus cyaneus*. As a boy Tristram found nests of the species close to his home in Northumberland. Now it is almost extinct as a breeding bird in England but is holding its own in Wales and Scotland.

Plate 2: The Golden Plover *Pluvialis apricaria* has suffered a small decline in recent years but is still to be found in many of its old moorland haunts.

Plate 3: Dotterel *Charadrius morinellus* can be found breeding on high land in Scotland and occasionally in its old haunts in the north of England.

Plate 4: The Gannet *Morus bassanus* has expanded its range in Britain over the last century and has established colonies on the north-eastern cliffs of Britain.

Plate 5: The Fulmar Petrel *Fulmarus glacialis* now breeds in almost all suitable sites around the British Isles. In Tristram's time there were none on the Farne Islands where now there are regularly almost 250 breeding pairs.

Plate 6: Audubon's Shearwater *Puffinus lherminieri* is mainly a diurnal bird, which was confused for many years with the Bermuda Petrel which is mainly nocturnal.

Plate 7: Bermuda Petrel *Pterodroma cahow* breeds only in the Bermuda Islands and is now recovering, as a result of conservation measures, from near extinction. Note the false entrance to the nesting burrow on the right which prevents access to the burrow for the competing Tropic Birds. Image by Jeremy Madeiros and published by permission of the Bermuda Conservation Services.

**Plate 8:** Green Heron *Buteroides striatus*, a species thought to be introduced to Bermuda, is now an established breeder.

Plate 9: Hudsonian Curlew (Whimbrel) *Numenius phaeopus hudsonicus* is treated as the North American race of the Whimbrel in del Hoyo et al., (1992–2013), though there is some evidence that it should be given specific status.

Plate 10: Red-winged Blackbird *Agelaius phoenicius*, a common species collected by Tristram in North America on his way home from Bermuda.

— 3 —

# Castle Eden to Canon: Birdwatcher to Ornithologist

LITTLE OWL. (*Athene persica.*)

Back in England Tristram was destined to make his home for the rest of his life in County Durham. However, this did not prevent him leaving the country on his numerous trips abroad and foreign expeditions, of which he was so fond, but he was to make his base for this in the North-East of England, closely associated for most of this time with one of the grand cathedrals of the world. There can be no doubt that by this time he was well on the way to becoming much more than a general naturalist pursuing a hobby, and with the contacts he had already made in Northumberland and Bermuda (see Chapter 4), the stage was set for his alternative career as an ornithologist.

No scientific venture is complete without the publication of the results and this had begun by his supplying Jardine with details of his observations in Bermuda. However, he had not yet entered into print on his own behalf, as any good scientist must, but this was now to change.

Immediately on his return to England from Bermuda, Tristram made his way directly to Durham City, to visit his step-mother and sisters. They then travelled with him to the village of Castle Eden, already having planned to make the visit before they knew of his return. Here he was able to accept in person the living of rector of that parish. The installation and induction at Castle Eden rapidly followed. Then, temporarily leaving the parish in the charge of another for a short time, he went to visit his friend Henry Bowlby in Whitby, specifically to enquire about the whereabouts of Henry's sisters. Finding that they were in Harrogate, and not being one to let grass grow under his feet, he promptly deserted his friend, and together with his portmanteau and two cases of butterflies, he headed for Yorkshire.

After arriving in Harrogate, however, he discovered that his portmanteau had not done so, but he still had his butterflies! After purchasing some new clothes Tristram visited his old friends, of which Louisa Tristram subsequently wrote: "After meeting with some difficulty (not from the lady), and after various obstacles had been overcome, to which it is needless here to refer, he was eventually married to the second daughter Eleanor Mary Bowlby on Feb.5th, 1850." (Tristram, L. H-H., 1898)

Louisa's report continues: "They spent their honeymoon in London and there the first place to which my mother was taken was a fusty bird shop where my father bought a Bulwer's Petrel." This specimen is probably the one recorded in Tristram's Catalogue (1889) as "*Bulweria columbina*, Madena, 1849—Dr Frere", now *Bulweria bulwerii*. After London the couple visited Oxford and Brighton and then stayed with the Rev. Edward Bowlby in Essex.

Back in the vicarage in Castle Eden they were in the environment of a country village but close to the industrial landscape of County Durham, which at that time surrounded it but has now encroached upon it. The description of the county, given by Tristram (1905), from an ornithological point of view clearly shows that he found it inferior to Northumberland. Of the coastline he wrote: "There is no shelter and little inducement for passing sea-fowl to halt on our coast" and "when we leave the coast the collieries and coke ovens which stud two-thirds of the county, destroying by their fumes the trees and hedgerows, and bringing a vast population, have in many ways driven away all the winged inhabitants save the house sparrow." However, this is merely describing the situation as it was then, for Tristram was no pessimist.

One of Tristram's favourite haunts in later life was Castle Eden Dene, which is now an SSSI (Site of Special Scientific Interest) and an NNR (National Nature Reserve). There is a series of denes running down to the sea on the coasts of Durham and Northumberland—deep ravines cut into the Magnesian limestone by streams running into the North Sea. Because of the steep sides of the ravines the woodland has not been felled and is regarded as semi-natural, typical ancient woodland with an extensive flora. Despite the fact that the dene, known locally as 'The Jungle', attracts some 15,000 visitors a year (a fact which Tristram would not like), almost two hundred species of birds have been recorded within the reserve, including nine species of warblers. Castle Eden Dene now contains the largest area of ancient woodland in the North-East and boasts many ancient trees—ancient even in Tristram's time. At that time the area would have had few visitors and Tristram would have been free to roam and indulge in his collecting activities, undisturbed. The ravine and its associated habitat are the results of glaciations in the last Ice Age which ended some twelve thousand years ago. Towards the sea the ivy-clad limestone cliffs, rising to a height of some thirty metres, give way to scrubland of blackthorn and hawthorn, and, on the higher ground, Magnesian limestone grassland, which is now a rare habitat.

The mouth of the dene opens onto dunes covered by marram grass, and a pebbly beach gives way to sand which extends several miles in each direction. In Tristram's time the beach would have been blackened by coal and not the attractive expanse that it is now. There are also other denes in the North-East which provide sanctuary for birds. One such is Jesmond Dene, just east of Newcastle-upon-Tyne, which has its own bird book, *The Birds of Jesmond Dene* (1931) by George Noble.

In the south of County Durham the Tees estuary afforded suitable and attractive habitat for seabirds and waders. In fact, the middle and lower reaches of the Tees, west of Darlington, provided excellent habitat for bird life and still do so. Once settled in the village Tristram found many good bird sites within the County, where, apart from when he was on one of his numerous excursions, he was to spend most of the second half of his life. This of course did not preclude his venturing back into Northumberland, to some of his favourite places such as the Farne Islands. He had ceased to keep a journal by the time he came to live permanently in County Durham so that the record of his activities remains only in the catalogues of his collections and in the unpublished accounts of his life written by his daughter Louisa and grand-daughter Eleanor. However, it is possible to derive some information from the correspondence with several of his ornithological friends. In his catalogues Tristram did not always record the date nor who collected the birds or eggs, but it is likely that those taken within the counties of Durham and Northumberland were taken in his presence even if his name was not appended to the specimen; for example, those taken by J. H. Gurney Jr (Tristram's pupil) on trips to the Farnes with its Shags (Plate 11), Kittiwakes, Puffins and Guillemots, and Coquet Island with its Eider Ducks (Plate 12).

Whilst Tristram collected extensively on his expeditions abroad, his collecting at home appeared to cover relatively few species and not many skins of each. Possibly he felt it politic to collect out of the sound of gunshot of his parishioners! There is no doubt that he exchanged many skins with collectors world-wide, so that it is possible that his Catalogue gives no real indication of the numbers of individual birds, or even the number of species that he shot. Similarly, the Catalogue of his egg collection shows that he took few of the eggs himself, even of the common birds in Durham and Northumberland. However, the catalogues do give an indication of the places which he visited.

Considering that his first collection of bird skins, now in Liverpool World Museum, consisted of over 17,000 specimens. Castle Eden got off lightly.

In eleven years residence there his Catalogue (Tristram, 1889) shows only some seventy birds were collected in the area, as many of the birds in his collection resulted from exchanges for foreign specimens that Tristram had himself collected. The seventy specimens were of thirty-six species, only eleven of which were Non-Passerines: Herring Gull (1), Common Gull (2), Lesser-black-backed Gull (2), Great Black-backed Gull (2), Kittiwake (1), Lapwing (1), Water Rail (1), Grey Partridge (3), Scaup (3), Kestrel (2) and Short-eared Owl (1). The Passerines were: White's Thrush (1), Redwing (4), Fieldfare (2), Song Thrush (3), Mistle Thrush (1), Garden Warbler (1), Willow Warbler (11) (Plate 13), Sedge Warbler (2), Dunnock (1), Coal Tit (1), Marsh Tit (3), Long-tailed Tit (2), Wren (1), Yellow Wagtail (1), Pied Wagtail (3), Tree Pipit (2), Pied Flycatcher (1), Spotted Flycatcher (1), Crossbill (2), Skylark (2), Starling (1), Hooded Crow (1), Carrion Crow (1), Magpie (1), and Jay (2) (Plate 14).

Away from Castle Eden Tristram's main haunts were the three main river valleys, the Tees, Wear and Tyne, the moorlands to the west, the estuary of the Tees and the Durham shoreline. The headwaters of all three of the main rivers rise on the eastern slopes of the Pennines, the Tees beyond the Durham border, under Cross Fell, the Wear in Durham under the watershed dividing it from the south Tyne which, like the Tees rises in Cumberland. All are within a distance of about six miles of each other. East of Blaydon-on-Tyne, the A19 near Sunderland and Thornaby-on-Tees in the south, the rivers become, as Temperley (1951) put it, thoroughly industrialised. In this area there was a great deal of industrial contamination but westwards there were vast areas of unspoiled countryside, even in Tristram's time.

The moors where the rivers rise are bleak places in winter, with often only a few grouse and carrion crows to be seen. In summer, it is a different story, with Curlew, Redshank, Common Sandpiper, Golden Plover, Dipper, Grey Wagtail and Meadow Pipit breeding commonly, and Merlin, Black Grouse and Cuckoo also present as breeders, but now uncommon. In wetter areas Teal and Snipe were to be found and, whilst according to Tristram (1905) the Black-headed Gull "has no breeding place left in the county"—meaning Durham—it still bred in 2013 in rather impermanent gulleries.

In the lower reaches of the rivers life flourished more in the valley of the Tees than in the other valleys, as it does to this day. Peregrine Falcon, Common Buzzard and Raven disappeared as breeding birds during Tristram's lifetime, which must have been a disappointment to him. In his summary of the county avifauna (Tristram, 1905) he was rather dismissive of the Durham coastline and estuaries as habitats for birds but he still visited what he termed "Teesmouth" where he collected a single Slavonian Grebe in the winter of 1869, a Little Grebe in the winter of 1862 and two female Oystercatchers in the autumn of 1867. In fact, his second living, Greatham, was adjacent to Cowpen Marsh and only a short distance from Seal Sands, so that in his time there (1860–73) he must have visited both areas regularly, and certainly did so on occasion whilst living at both Castle Eden and Durham.

That the then Tees Estuary retained much of ornithological interest was shown by the results of the Coatham Marsh Decoy, which was constructed on the Yorkshire side of the estuary, and duck were caught in large numbers. The area enclosed by the decoy was of some three acres with a pool the size of two acres, which increased in size when the marsh was flooded. Temperley (1951) records that catches included Mallard, Wigeon, Shoveler, Pintail, Teal, Pochard and Shelduck. On one occasion nearly five hundred duck were enclosed "in the net", which unfortunately (but not for the duck!) broke due to the weight and only one hundred were caught. Payne-Gallwey (1886) records this event rather differently: "as many as one hundred ducks and Wigeon having been taken at once in a single pipe; though on this occasion owing to the weight of fowl enclosed, the net broke and the greater number escaped." Despite the differing accounts it is clear that duck were common on the Tees estuary, at least at the beginning of the life of the decoy. The large catches were recorded, according to Payne-Gallwey, around 1860, about the time when Tristram moved from Castle Eden to Greatham. His appointment was as vicar of Greatham and Master of Greatham Hospital, and during his time there the industrialisation of the area continued apace. As a result of this the catches there decreased and his departure from Greatham more or less coincided with the building of the blast furnaces near the decoy and its subsequent demise.

In his thirteen years in Greatham Tristram collected 67 birds of 32 species and in Castle Eden he had taken 70 birds of 36 species. Elsewhere in County Durham, outside his two parishes, he obtained 77 specimens of 28 species over a period of forty years, so that throughout the county he collected 214 specimens which were recorded in Tristram's Catalogue (1889). In his two parishes he apparently obtained 137 birds in twenty-three years, approximately 6 birds a year, most of which he probably shot well away from the village and outside his parishioners' hearing.

In his account of the birds of County Durham (Tristram, 1905) he records a total of 247 species of which he personally collected representatives of 88, approximately one third of the total. Nearly all of these could reasonably be described as common birds, as only two of his acquisitions were rarities. The White's Thrush *Zoothera dauma* was the eighth record for Britain at the time, taken 31 January 1872, when Tristram was still in Greatham. It was originally shot and wounded by Mr Rowland Burdon a fortnight before its capture and it lived for a further three weeks after this. The specimen was given to Tristram by Mr Burdon. Of more biological significance was the Pallas' Sandgrouse, a male of which species Tristram shot in Greatham on 20 May 1863. Recorded first in Britain in 1859, it was unrecorded in Durham until 1863. "From May to July many were seen and taken on the coast, on the sandhills of Seaton and on Cowpen marshes." (Tristram, 1905) Such "irruptions" as these occurrences are termed, occur from time to time when birds are especially numerous or in imbalance with their food supply (Newton, I., 1985), and are important from the point of view of a possible resulting range extension of the species concerned. This irruption of Pallas' Sandgrouse, which Tristram observed, was one of the first to be recognised as possibly being a method by which the future distribution of a species might be affected. After a later irruption in 1888, the sandgrouse bred in Yorkshire but failed to establish a breeding population. The Common Crossbill, which does breed in Britain, is the most irruptive species which visits our islands at relatively frequent intervals, and such an irruption in the past may have given rise to the breeding population of the Scottish Crossbill, now recognised as a distinct species, and later to the Common Crossbill further south.

Alfred Newton (1864) published a very extensive account of the 1833 irruption of the Pallas' Sandgrouse and Tristram had his letter to *The Times* newspaper republished in the *Zoologist* (1863), in which he made no mention of the bird that he had acquired but recorded that all the specimens obtained were males.

Tristram's residence in County Durham can be conveniently divided into three periods, firstly as vicar of Castle Eden (eleven years), secondly as vicar of Greatham and Master of Greatham Hospital (thirteen years) and thirdly as Canon of Durham Cathedral (thirty-three years). In this last period he resided in the city in 'The College', a square off the South Bailey. It was during the first of these periods that he began his writing and his exploration of North Africa and Palestine. At the age of thirty-one Tristram published his first ornithological note, "Occurrence of Little Auk (*Uria Alle*) in the city of Durham", (Tristram, 1853a). No-one could have predicted the literary avalanche which followed.

In the summer of 1852 Tristram visited Denmark and Norway. No record of the visit appears to be in existence except that which is included in the Catalogue (Tristram, 1889). In Aarhus, Denmark, in July of that year, he collected three specimens of the Long-tailed Tit *Aegithalos caudatus caudatus*, the Scandinavian white-headed form of the species. Further north, in Norway, he collected specimens of the Black-throated Diver, Great Black-backed Gull and Bar-tailed Godwit at Bodo, Spotted Redshank, Ptarmigan and a Three-toed Woodpecker at Kjerringoy, and a Black Woodpecker listed only as "Norway 1825". He also purchased two Capercaillie which he added to his collection.

In the spring of 1855 Tristram again became ill and travelled to Torquay with his step-mother rather than his wife, who at the time was heavily pregnant, where they spent six weeks with his uncle and aunt, the Rev. George and Mrs Almond (Fleming, Eleanor, n.d.). Whilst in Torquay, Tristram wrote to Newton, still addressing him as "Dear Sir," as he had done previously, asking that he obtain a Brambling in summer plumage for his collection and informing him that "I am happy to say that I am deriving much benefit from the air of Torquay." (Letter 9839/1T/204) Following this, and on the advice of his London medical advisers, it was decided that he should spend the following winter abroad. Algiers was decided upon as the destination for the winter sojourn, and his wife and his newly-born daughter, together with two of his pupils, were to accompany him. His medical adviser gave his wife to understand that she could never expect to bring him home again. Clearly Tristram disagreed with this diagnosis and on arrival in Algiers quickly got well again, and, liking the country, he decided to arrange an expedition into the centre of Algeria in the following winter of 1856–7 (see Chapter 5).

Tristram had now begun a family but by his action in taking his wife and Mary to Algeria he was clearly not going to allow this development to interfere too much with his other activities. Over the next fourteen years, seven more children were to be born. Following Mary (1850–1932) were Louisa (1852–1907), Eleanor (1853–1946), Charlotte (1855–1938), Katherine (1858–1948), Christiana (1860–1954), Henry (1861–1946) and Frances Anne (1864–1944). Mary married the Rev. William Lyall Holland, Eleanor married Hely Hutchinson Almond, Christiana married Percy Heawood, and Henry married Emmeline Worrall. Louisa, Charlotte, Katherine and Frances were unmarried.

The children grew up with Tristram's collections and according to his granddaughter Eleanor (Fleming, E. M., n.d.) his daughters "did not appreciate the amount of room and the dusting entailed", and an aunt, one of the daughters involved, on being asked in what branch of science she was interested replied, "Astronomy, because you can't collect stars!" Even Henry was unenthusiastic about Tristram's pursuits. However, they all seemed to enjoy the many visitors to the house, particularly after Tristram's return from his second and major expedition to Palestine in 1864 (Tristram, L. H-H., 1898). Of these Alfred and Edward Newton Louisa dubbed "facile princeps" with Alfred Russel Wallace a close runner-up. Louisa describes Alfred Newton and Alfred Wallace entertaining the little girls by carving apples and building respectively a pit-head gear and a swan with spread wings.

In all Tristram's travels Louisa stresses that "the interest of his little girls was never lost sight of. He entered into all their pleasures and tried hard to imbue them with his own love of natural history . . . for the most-part not very successfully. . . . He shared all his interests with his children as soon as they were in any way able to understand them." When he was at home Tristram was obviously a very attentive father and Louisa wrote enthusiastically about the time when he took his four eldest girls to the Exhibition in London and showed them the sights. The greatest treats, however, were riding on the top of the omnibus "when it was considered quite impossible for a lady to do such a thing".

Henry Barrington Tristram, the only boy in the family, was educated at Loretto School, Winchester College and Hertford College, Oxford, where he read Classics. He played rugby for England and cricket for Oxford University and Durham County Cricket Club. Subsequently he was Headmaster of Loretto School, St Paul's School and later moved to Jersey where he was Head of Victoria College. Katherine became a missionary and Louisa, at the age of 18, on leaving school, became her father's secretary. She writes: "I also had the privilege of arranging his eggs though I could never feel the affection that a naturalist's daughter ought to feel for her father's collections. I much preferred the writing." (Tristram, L. H-H., 1898)

Whilst still residing in Castle Eden, in 1858 Tristram took a winter cruise with friends into the Mediterranean and spent some three weeks in Palestine, the first of six visits. He returned to Castle Eden in the late summer of 1858, having formed a close friendship with Bishop Govar of Jerusalem. He had collected several new birds, despite having stayed in Palestine for such a short time and resolved to return and explore the country more extensively in the future in so far as he was permitted. Tristram remained at home in Castle Eden, much revived by his travelling, until the beginning of 1860 when he moved to the living of the parish of Greatham, close to the Tees marshes.

In the previous few years, Tristram had begun writing seriously, first short notes to the *Zoologist* concerning a Little Auk (*Uria alle*) found dead in the City of Durham in 1853, (Tristram, H. B., 1853a) and secondly, on the occurrence of a Jacamar (*Galbula ruficauda*) in Lincolnshire in 1859 (Tristram, H. B., 1853b).

In the first volume of *Ibis* (1859a), the journal of the newly-formed British Ornithologists' Union, he contributed his first paper on the birds of Palestine (the second paper listed in the first number of the new journal).

In addition two papers on the birds of North Africa (the second in three parts) also appeared in the first volume of *Ibis* and his magnum opus *The Great Sahara* was published the same year. He also became President of the Tyneside Naturalists' Field Club in 1860 and his Presidential Address was published in its Transactions.

Tristram was to remain vicar of Greatham for the next thirteen years during which time he visited Palestine on three occasions (Plate 15). The second of these was his epic ten-month expedition of 1863–4 where he was frustrated by being prevented from exploring to the east of the Dead Sea (see Chapters 8 and 9), and the third in which he achieved this ambition (see Chapter 10). Between these two he published three more important works, *The Land of Israel* (1865c), *The Natural History of the Bible* (1867b) and *The Topography of the Holy Land* (1871). In addition he published extensively on his ornithological interests in the *Ibis* and these are shown later in his list of publications.

In 1870 Tristram was appointed an Honorary Canon of Durham Cathedral and whilst this did not increase his income, neither did it affect his non-clerical activities. This changed in 1873 when the Bishop of Durham offered him the vacant residential canonry on the death of the Rev. Temple Chevallier who was an astronomer, mathematician and Professor of the University (Baker, 1996). On the appointment the Bishop required him to give up the living of the Parish of Greatham and to take up residence in the city. About the latter Tristram was unenthusiastic but on finding again that his extramural interests would be unaffected, he accepted the appointment and moved into the College, off South Bailey, in 1874.

Soon Tristram became Honorary Curator of the University Museum and this resulted in closer contact with his old friend William Proctor, the Sub-Curator. He obtained much new material for the Museum, supplying quite a lot himself. Because of the damp in the old fulling mill on the river he transferred the more delicate items, such as mounted birds, to the alms houses on Palace Green. In 1877 William Proctor died. He had become a very competent natural historian over the years and would have been difficult to replace. In the event Tristram consulted his friend John Hancock who had many contacts in the North-East (Letters NEWHN:1967:H67:1265 &

1266) and eventually Joseph Cullingford was appointed as Sub-Curator. Six years later the Museum was at its best and Tristram wrote to his fellow ornithologist Gunther in the British Museum stating that it was "a regularly organised Museum with a secured trust in perpetuity". (NHM, Keeper's Archives 1.24:392) This was only the second university museum to be open to the public. As far as the Museum was concerned "perpetuity" did not last much longer than Tristram, for in 1917 the natural history specimens were dispersed, many of the birds being taken into St Hild's College, many others being thrown away. Someone appreciated the value of the Great Auk which was rescued and it spent the duration of the First World War in a box "in the wardrobe of the wife of a university caretaker". (Baker, 1999) After the War it was sent to Jamrach's in London for refurbishment and was put on display in the library on Palace Green until 1924 when it was transferred to the Dawson Building on South Road. There it remained until 1946 when it made a small migration within the building to the office of the Head of Department of Zoology, where it graced his filing cabinet and the supervision of numerous research students.

In the meantime the old fulling mill had become the Archaeology Department and then the Archaeology Museum, when the Department got new accommodation. The Great Auk remained protected until the departure of Professor J. B. Cragg in 1962. Unprotected and unappreciated by the powers that were, the Durham Great Auk became the Glasgow Great Auk. In 1986 a permanent exhibition of Archaeology was opened in the old fulling mill; probably the Great Auk would have felt at home here as it is now only through archaeological excavations that any further knowledge of the species will be obtained. The price of £30,000 paid for the Great Auk by the Glasgow Museum authorities caused debate within the pages of the journal *British Birds* on how such a sum might be better spent on the conservation of extant but threatened species. Under the headline 'Great Auk good buy', chronicler of the Garefowl Errol Fuller made the very significant point: "The stuffed Great Auk should captivate the minds of Glaswegians for many years to come" (Fuller, 1995), and thus be a lasting reminder of what did happen and that a recurrence of another such happening might be prevented by suitable action in the future.

— 4 —

# An Exaltation of Ornithologists

SACRED IBIS. (*Ibis religiosa.*)

In 1853, shortly after taking his B.A. degree, Alfred Newton was appointed to the Drury travelling fellowship for the sons of Norfolk gentlemen, at Magdalene College in the University of Cambridge. This appointment was pivotal in the development of the subject of ornithology, not only in Great Britain but in a much wider sphere. Initially Newton travelled widely, paying much attention to the Arctic, but his returns to Cambridge were frequent so that he had opportunities to strike up new contacts at home as well as abroad.

Amongst these new contacts were three undergraduate students who entered the university about the same time as Newton's younger brother Edward, and it was probably through him that Alfred Newton was first introduced to them. All four had a strong interest in natural history, and particularly in birds, and from time to time they met together in Newton's rooms in Magdalene to discuss their ornithological interests. Of these Edward Newton was the eldest, being born in 1832, whilst Fredrick DuCane Godman was born in 1834, Osbert Salvin in 1835 and Percy Sandon Godman in 1836. Two others were involved in these meetings, Henry Maurice Drummond and Henry Baker Tristram who had spent time together in Bermuda.

Tristram had married Eleanor Mary Bowlby in 1850 and she was cousin to Osbert Salvin who may well have been involved in introducing Tristram to the group, which from time to time included others, particularly other close associates of Newton such as Philip Lutley Sclater, John Henry Gurney and John Wolley. Amongst the topics discussed from time to time at the informal meetings of this group, of which the undoubted leader was Alfred Newton, were the possibilities of establishing a more formal and more extensive Society and a regularly published Journal. However, such was the informality of these meetings that no records were ever taken and information concerning them existed only in the memories of the participants.

Where there are no written records of a happening it is often difficult, or impossible, to determine when an event took place or even who was present. Even if there are records, is the time of the first airing of an idea of less importance than the timing of decisive action? In the Preface to the first volume of the *Ibis*, dated October 1859, the first editor of the journal, Philip Lutley Sclater, records:

> For some years past a few gentlemen attached to the study of ornithology, most of them more or less intimately connected with the University of Cambridge, had been in the habit of meeting together, once a year, or oftener, to exhibit to one another the various objects of interest which had occurred to them, and to talk over former and future plans of adding to their knowledge of this branch of natural history. In November 1858, the assemblage took place in Cambridge, and after due consideration, it was determined by those present that a Quarterly Magazine of General Ornithology should be established, that a limited subscription should be entered into to provide a fund for that purpose and that the subscribers should form an 'Ornithological Union', their number at present not to exceed twenty.

Unsurprisingly, Alfred Newton, not yet then Professor, was more concerned about who were the founder members and in a letter (in Wollaston, 1921) to Tristram dated 2 January 1888, listed those present at the Conference in his rooms in November 1858, as "yourself (Tristram), Drummond, P. L. Sclater, and E. C. Taylor whom he brought down, the Godmans, Salvin, Sealy, Simpson, and A. [Newton himself] and E. N. [Newton's brother]". However, he did not recognise all these as founder members! "Don't forget that E. N. [Edward Newton] was emphatically one of the founder members of the B.O.U., which is a good deal more than being only one of the original 20. I have always looked on the founders as:

- Drummond
- Tristram
- Newtons (2)
- Salvin
- Godmans (2)

The rest—Sclater, Gurney and Wolley included—were asked to join us". (Fig. 1).

Ibis. Jub. Suppl., 1908.

*Ornithological Union of England*

Robert Birkbeck, Esqr., F.Z.S. 65 Lombard Street.
Lieut. Col. H. M. Drummond, (late 42nd Royal Highlanders)
  Megginch Castle, Perthshire. President.
J. C. Eyton, Esqr., F.L.S., F.Z.S., &c. Eyton Hall, Wellington, Shropshire.
Fred. Godman, Esqr., M.R.I., 55 Lowndes Square.
Percy Godman, Esqr., B.A., Park Hatch, Surrey. Borregard, Sarpsborg, Norway.
J. H. Gurney, Esqr., M.P., F.Z.S., Pres. Norf. & Norw. Mus., Catton Hall, Norwich, Norfolk.
Rev. W. H. Hawker, M.A., Horndean, Hants. Dickmox Esqr. Midhurst, Sussex –
Edw. Clough Newcome, Esqr. Hockwold Hall, Norfolk.
Alfred Newton, Esqr., M.A., Fellow of Magdalene College, Cambridge,
  F.L.S., F.C.P.S., F. Sch. Lit. S.
Edward Newton, Esqr., B.A., Elveden Hall, Suffolk.
Capt. Poulett Orde (late 42nd Royal Highlanders)
Honble. T. Lyttleton Powys, F.Z.S., Lilford Hall, Northants.
  Salvin Esqr. Corr. Memb. Zool. Soc.,
Osbert Salvin, Esqr., B.A., 11 Hanover Terrace, Regents Park.
Philip Lutley Sclater, M.A., Fellow of Corpus Christi College, Oxford,
  F.L.S., F.Z.S., &c. &c. Editor of "The Ibis".
A. F. Sealey, Esqr., M.A., 70 Trumpington St., Cambridge.
W. H. Simpson, Esqr., M.A., 21 Gloucester Place.
Rev. E. C. Taylor, M.A.,
Rev. H. B. Tristram, M.A., F.L.S., Rector of Castle Eden, Durham.
John Wolley Jr., Esqr., M.A., F.Z.S., F. Sch. Lit. S., Beeston, Nottinghamshire.

Figure 1: Alfred Newton's list of the Original Members of the British Ornithologists' Union. *Ibis* Jubilee Supplement, 1908, Vol. 50. Reproduced with the permission of the British Ornithologists' Union, <http://www.bou.org.uk>.

Of the older members of this group the then Col. H. M. Drummond was mainly out of Europe until after he resigned his commission in 1852, making the famous last sighting of a live Great Auk on his return journey, off the Newfoundland banks. In 1853 he was back in Perthshire but effectively isolated from the Cambridge group. Further south Tristram was also back in Britain, at Castle Eden, in County Durham, where he had been in residence since 1849. The rest of the group were still students in the early 1850s so initially it was more likely that Newton would form a friendship with Tristram at this time. This he did, and whilst the first recorded correspondence between the two was not until 1854, it is quite likely that the two had made contact sometime before this as a result of Tristram's relationship with Salvin, already mentioned. Tristram and Newton became firm friends and remained so until Tristram's death in 1906.

Newton's "Conferences", held in his rooms in Cambridge, probably first took place in either 1854 or 1855, and it may well have been as early as this that the possibility of an ornithological society and learned journal were first mooted. However, the first record (Mountfort, 1959) of any such meetings concerned two in 1857, the first at Tristram's house in Castle Eden and the second at the Bull Hotel in Cambridge. In her 'Recollections' Louisa Tristram writes of a summer meeting in 1858 "where friends Mr Wolley, Mr Alfred Newton and Mr Osbert Salvin met in my father's study, and together they planned the publication of a first class ornithological magazine." (Tristram, L. H-H., 1898) At the British Association meeting in Leeds, in September 1858, further discussion took place on these matters and the final decision came about in Newton's rooms on 17 November 1858 when both the *Ibis* and the Union were launched. Sharpe (1907), in his Presidential Address to the Fourth International Ornithological Congress, asserted that the first planning meeting occurred in Canon Tristram's house in Castle Eden. No matter which location, nor the timing, it is clear that Tristram played a central role in the founding of both the Union and the journal.

The immediate priority, once the decisions were taken, was to produce the first volume of the journal and there was some debate about the title (Fig. 2). 'IBIS' was the name suggested by Dr William Francis of the printers, Messrs Taylor and Francis. Sclater (1909), the first editor, immediately adopted the idea, and since Newton was also pleased, the name was more or less accepted. However, John Wolley was much against the idea and argued strongly for the name 'Avis'. A hundred years later in a volume of *Ibis* celebrating the centenary of the BOU, the then editor wrote: "There was much controversy about the name the journal should bear and one cannot help regretting that the suggestion of 'Avis;' was overruled. There is so much to be said in favour of the universality of this name that one wonders why it was not accepted." (Moreau, 1959) In writing to Tristram on 10 December 1858 (Wollaston, 1921), during the period when the first volume of the *Ibis* was being prepared, Newton wrote: "I hope things are promising for the Ibis; we nearly lost Wolley through the change of name, but I trust he is appeased."

There was almost as much controversy about the motto to be printed on the title page (Fig. 2) as there had been about the title of the journal. According to Moreau (1959) these lines were fabricated by the editor, P. L. Sclater:

> Ibimus indomiti venerantes Ibida sacram
> Ibimus incolumes qua prior Ibis est

and translate as follows:

> Venerating the Sacred Ibis we shall go fearlessly
> Safely we shall go in Its presence

Newton was certainly critical: "The Ibis motto I confess I do not understand." Whether anyone else did went unrecorded. When Newton took over the editorship he changed the motto, and thereafter there were numerous changes until in 1946 it was dropped altogether. There appears to be no record of Tristram having commented on either the name of the journal or the motto, but he would probably have approved of the cover designed by Joseph Wolf depicting a Sacred Ibis standing amongst ruins in ancient Egypt. This survived until 1946 when the sun set literally on the scene and two years later Wolf's image was eclipsed by the members voting in C. F.

# THE IBIS,

## A MAGAZINE OF GENERAL ORNITHOLOGY.

EDITED BY

PHILIP LUTLEY SCLATER, M.A.,
FELLOW OF CORPUS CHRISTI COLLEGE, OXFORD;
SECRETARY TO THE ZOOLOGICAL SOCIETY OF LONDON;
FELLOW OF THE LINNEAN SOCIETY; HONORARY MEMBER OF THE ACADEMY OF NATURAL
SCIENCES OF PHILADELPHIA, OF THE LYCEUM OF NATURAL HISTORY OF NEW YORK,
AND OF THE GERMAN ORNITHOLOGISTS' SOCIETY; ETC.

VOL. I.   1859.

"Ibimus indomiti venerantes Ibida sacram.
Ibimus incolumes qua prior Ibis adest."

LONDON:
N. TRÜBNER AND CO., PATERNOSTER ROW

| Paris. | Leipzig. | New York. |
| Fr. Klincksieck, | F. A. Brockhaus. | John Wiley, |
| 11, Rue de Lille. | | 56, Walker Street. |

1859.

Figure 2: The Title Page of the first volume of the Ibis, 1859, Vol. 1. Reproduced with the permission of the British Ornithologists' Union, <http://www.bou.org.uk>.

Tunnicliffe's much simplified image. That the Sacred Ibis (Plate 16) remained on the somewhat simplified title page would have pleased Tristram but he would have preferred the original, behind which was bound one of his most important papers—Tristram (1859c), Part III of his third paper to appear in the first volume of *Ibis*.

Moreau (1959) remarks that "It is noteworthy that the foundation of a journal was the original inspiration of our society, not, as is perhaps more usual, the other way about." Having founded the journal it was necessary to provide financial support for its foundation and continuance. The arrangements for this were left to Sclater and Newton and together they came up with the list shown in Fig 1. The list itself is in Newton's hand and annotated by Sclater. Newton was to be Secretary, Sclater the editor of *Ibis* and Drummond was elected President. Since this was decided at the meeting on 17 November 1858, this should really be the date of foundation of the Union, the first Annual General Meeting being held on 9 November 1859. The first number of the *Ibis* appeared in March 1859 and the Jubilee Meeting was held in March 1909.

It is ironical that the membership list headed erroneously by Newton as 'The Ornithological Union of England' should have had the only Scotsman amongst them elected as President. This was the first membership list of the British Ornithologists' Union and the group which had previously met in Castle Eden and Cambridge formed the core of it. At the end of April 1859, as the second part of the first volume of *Ibis* was issued, its editor Philip Lutley Sclater was supposedly selected by Richard Owen and William Yarrell, and subsequently appointed to be secretary of the Zoological Society of London. Yarrell probably had no involvement in this as he died in 1856. In 1861 Sclater was elected as a Fellow of the Royal Society (FRS). On 4 June 1868 Tristram was similarly honoured, with Charles Darwin as one of the signatories of his certificate of proposal. Thus within ten years of the founding of the BOU three of the original members were Fellows of the Royal Society and it could be argued that Tristram was the first to be elected solely on his contributions to ornithology.

On 20 November 1859 John Wolley died and his place was taken amongst the remaining nineteen members by Mr Robert Fisher Tomes, a Corresponding Member of the Zoological Society. With the formation of the Union, Tristram's circle of ornithological contacts was expanding and he made further important contacts by the decision taken to elect ten ornithologists not resident in the United Kingdom as honorary members. The decision to continue to limit the number of ordinary members to twenty was surprising since these members were supporting the publication of the *Ibis* but equally surprising were those ornithologists outside the union, such as Jardine and Selby.

Fourteen and a half of the thirty-six papers which appeared in the first volume of *Ibis* in 1859 were written by Salvin, Newton and Tristram. John Wolley had submitted one before his death, and the Godmans and Drummond did not offer anything. Sclater solicited these papers from the authors, wrote two himself and as editor perhaps made the greatest contribution. For many years to follow this group of friends formed the active centre of the Union.

By this time Sclater had joined the small group of original founder members, as defined by Newton, and edited the first six volumes of *Ibis*. He was a most competent administrator and as the new Secretary of the Zoological Society was very much involved with its Proceedings and the Transactions, which were several years in arrears. He continued to edit *Ibis* until the end of 1864 when he relinquished the post to Newton. Sclater remained as Secretary of the Zoological Society until October 1902 and he left it as a much larger and better organised institution than it was at the time of his appointment.

The small group that founded the BOU met frequently at scientific meetings and, apart from the Union meetings, often attended those of the Zoological Society and particularly those of the British Association. This was divided into different scientific sectors, and zoology was dealt with in Section D. Tristram made a great effort to attend these meetings, which he seemed to enjoy particularly and gave him an additional excuse, if this was necessary, to meet his ornithological friends.

Looking at photographs of Victorian gentlemen in general the impression is given of their really being rather severe, restrained and not easily amused. However, Tristram's granddaughter, Eleanor, remarked of her grandfather that "the fund of humour was always bubbling over" (Fleming, n.d.) and it would appear that this was not limited to his association with his grandchildren. (Plate 17)

Within the British Association was "a sort of Society or club" (Wollaston, 1921) called the Red Lions "whose object is convivial". The name was derived from Red Lion Court, Fleet Street, the address of Taylor and Francis, printers of the *Ibis*. (J. C. Collins, personal communication). Each meeting they assembled for dinner and invited a few guests from the meeting place who were specially connected with the Association. The Chairman was known as the King Lion, and his right-hand men, two "Jackals", would send out invitations indicating that "Bones will be provided..." at such-and-such a time and place. No serious talk was allowed and each Lion was required to be as humorous as possible. "After dinner comic songs were sung, ludicrous speeches delivered, burlesque lectures or papers given." (Wollaston, 1921) Approval was marked by a "roar", disapproval by a "growl". Telegrams were received from absent members, usually incomprehensible to those not knowing the individuals concerned. The following example was probably sent by Huxley whilst on a visit to the USA: "American Eagle now waving centennial wings greets Red Lion. How many of our scientists will we extradit in swap for Huxley who is having quite a nice time on this side and concludes to stop. Would M'Kindrick like Sitting Bull for vivisection? Wire reply, U.S. Grant. Telegram received and read to the Den by the Lion King, Glasgow, Sept. 11, 1876."

It would appear that such levity was not limited to Red Lion dinners as on one occasion Newton suggested to Mrs Strickland the title of a paper for the Zoological Section of the BA at Bristol: Sir John Lubbock to speak 'On the inability of Bees to avail themselves of Bank Holidays'. Newton appeared to have a penchant for such humour and even involved Tristram. At a meeting in Bristol, Louis Napoleon, himself a famous ornithologist, was expected and Newton and Tristram, awaiting the coming of their friends, witnessed the arrival of Sclater in Rowley's carriage—"a swell affair"! Descending, Rowley said, "There they are", meaning Newton and Tristram. The Mayor thinking that this meant the arrival of Louis Napoleon rushed across, thinking that Sclater was he, when "Sclater with great presence of mind presented Tristram as the Emperor, whereon the Mayor got furious and turned to me [Newton] with 'Who are these persons?'" (Letter to Mrs Strickland 13 June 1876; in Wollaston, 1921). (Plate 18)

Perhaps Newton was responsible for this apparently un-Victorian frivolity, which seemed also to run through his correspondence and even extended to the cover of reprints he sent out. A copy of one in the possession of the present writer reads: "Viro illustri primo reverendoque H. B. Tristram, D.D. aucto AN.". This dedication in many ways sums up the relationship between Newton and his great friend Tristram in that it contains five puns: "To a man (who is) Illustrious (P)rim(at)e and (R)evere(n)d". Newton also christened Tristram "Sacred Ibis", and, on his canonisation, "The Great Gun of Durham".

The tenor of these meetings of ornithologists reflects the tenor of their relationships, and Tristram and Newton were clearly very much amused by it all. Like birds, ornithologists lend themselves to collective nouns and no such noun is more applicable to this grouping of "Original Founders" of the BOU than 'Exaltation'. An Exaltation of Larks—an Exaltation of Ornithologists—a pun almost worthy of Newton and a personal reminder of David Lack's "Alauda vespertina".

Plate 11: Shag *Phalacrocorax aristotelis* is still a common breeder on the Farne Islands as it was in Tristram's time.

Plate 12: A male Eider Duck *Somateria mollissima* (St Cuthbert's Duck), still a common resident of the Farne Islands and with which Cuthbert is said to have "communed" regularly whilst he lived there.

Plate 13: Willow Warbler *Phylloscopus trochilus*, probably the commonest of the warblers present in Castle Eden during Tristram's residence there.

Plate 14: Eurasian Jay *Garrulus glandarius*, a common species in Castle Eden, which Tristram encountered in both North Africa and Palestine. He recognised the continental birds as a different species but today they are given only sub-specific status.

Plate 15: Henry Baker Tristram as he dressed during his explorations of North Africa and Palestine. Reproduced by the permission of St John's College, University of Durham, and Mr Robert Gladstone.

Plate 16: Sacred Ibis *Threskiornis aethiopicus*, the emblem of the British Ornithologists' Union.

Plate 17: Henry Baker Tristram during the late 1860s.

— 5 —

# The Sands of the Sahara: the Western Desert

LAMMER GEIER.  (*Gypaëtus barbatus.*)

Had Tristram been alive today he would happily have substituted his gun for a pair of binoculars. Out with Tristram during the 1850s, however, it would be no unusual event to hear the discharge of a firearm except, perhaps, if it happened to be aimed at you. About three hundred miles inland south of Algiers lies the town of Ghardaia. Some two days march short of the town, and having passed through the ravine of the Wed Ballough, Tristram's party was then again in the rocky Chebkha and pitched camp for the night. Soon after setting off next morning Tristram spotted a gazelle and rode off in pursuit with his guard Omar. In full stride Tristram "felt the pinge of a ball past my eyes and with it heard the report of my companion's gun . . . drawing my revolver I begged him to give me the flint from his gun . . . and he reluctantly surrendered." Tristram believed that this was an intentional act of attempted murder but apparently forgave his companion. This was but one of the problems that beset the Victorian ornithologist on his venture into the great Sahara; unreferenced quotations in this and the next chapter are taken from Tristram's first book, *The Great Sahara* (1860b).

Some weeks before this incident Tristram's small expedition had set out south from Algiers in the September of 1856. He had spent the previous winter in Algiers on medical advice and from there made several excursions into the interior. On this second expedition, this time accompanied by his friend the Rev. James Peed, "also in quest of health", they intended to spend this second winter "altogether in the Sahara". *The Great Sahara* documents these "wanderings south of the Atlas Mountains". It is a travelogue of topography and population, mostly a book of the desert and the people who occupy it, but it is also the first insight into the birds of this part of the African continent. He documents the birds which he shot and those which he just observed, firstly in Appendix V of the book and also in a series of papers in the *Ibis* (1859b; 1859c Parts 1, 2 and 3; 1860a Part 3 concluded; and 1860b). From being a reluctant writer on his experiences in Bermuda, Tristram had suddenly become an avid communicator!

On this second expedition Tristram had made his first documented observations as his boat was crossing the Mediterranean when "We were reminded that autumn was now setting in by the numerous flocks of migratory birds which passed us on their way southwards, chiefly familiar English summer visitors, whitethroats and warblers and many short-toed larks and pipits."

Most of the equipment for the expedition needed to be purchased in Algiers, and even in those times a good deal of red tape needed to be circumvented before the expedition could proceed. Certificates to carry guns and more valuable documentation such as letters to various generals and commandants of the interior, without which they could not have moved a step, were in fact less trouble. In letters provided by the Governor Tristram was described as "naturaliste très distinguée" (apparently much to Tristram's embarrassment) and Peed as "archaeologiste profonde".

At the end of the previous winter (1855–6) Tristram had visited Lake Halloula, near Blidah, which in those days was an ornithological paradise. There he had found the nests of over thirty species in a day in the spring of 1856. In the reeds he had seen some of the rarest of European warblers, and many ducks and herons were also present. It has since been drained and is now farmland. This would have been an excellent site for Tristram to have collected warblers, and his general account of his visit to the lake (Tristram, 1860a, Part IV) records that he did. However, he did not give exact dates for the locations he visited nor did he label many of the specimens with details of where they were collected, other than to write "Algiers" e.g., A pair of Montagu's Harriers, collected on April 12th 1856 (Plate 19). His Catalogue (Tristram, 1889) records only a male and female of the Subalpine Warbler as being taken at Halloula. Perhaps other specimens were given away or exchanged. Even so, his Catalogue reveals that during the two winters he spent in Algeria he collected twenty-two species of warbler.

In the spring of 1856 Tristram set out alone from Algiers with the main object of seeing Lake Halloula (Tristram, 1860a). His initial path lay through the forest which he described as being occupied by few birds except on its edges and in its few clearings. Here he obtained his first Algerian Jay and the eggs and nests of Nightingale and Algerian Greenfinch. Neither the latter nor the Jay is now regarded as specifically distinct from the European forms. Incredibly, he encountered an English farm worker, probably the only one in North Africa, and stayed for breakfast at the farm. Two families worked on the farm and the two wives had encountered two leopards a few days previously on one of the farm paths. Despite the fact that he explained to the women that for an unprovoked leopard to attack a human was "unheard of", Tristram had no wish to encounter a leopard at night so was soon on his way. His luck continued, for close to Lake Halloula he came across a camp of Zouave (French infantrymen) convicts, amongst whom was one who had an interest in birds and had worked in Paris for M. Verreaux, the French ornithologist. Tristram obtained his help in skinning some of his specimens and then "obtained a holiday for my new friend that he might accompany him in the morning. Before turning in, I spread in the camp amongst the convicts an announcement that for all nests brought me *with the bird snared and alive*, within the next three days, I would pay at the rate of one sou per egg."

The next morning Tristram's new convict companion immediately justified his hire by bringing down the first Subalpine Warbler he had seen. Savi's Warblers were common but had apparently not started nesting, but three nests of the "Melodious Willow Warbler, with complete clutches of four eggs, were found in the drier part of the thicket—very different in the position and texture of its nest from that of our Willow Wrens." Tristram described the nest as "extremely compact and neat, not unlike that of a Goldfinch . . . generally on the bare fork or branch of a tamarisk." Further searching of the grass revealed the nest of a Sardinian Warbler—a loose construction, very neat and round and "comfortably lined with fur and wool" and also that of the Cetti's Warbler with four brilliant red eggs "so strangely different from those of every other warbler." The depth of the nest was more than twice its diameter, "composed entirely of coarse grass outside and finer stems within but with no lining of hair or feathers."

Turning his attention from the nests and eggs to the birds themselves, Tristram was attracted to a singing bird which he did not immediately recognize. The normal procedure in those days was not merely to observe the bird, make notes and later consult a field guide—but to shoot it. This he did—what is hit is history, what is missed is mystery! The bird turned out to be an Olivaceous Warbler, then referred to as a Pallid Warbler *Sylvia elaica*, thought by some to be distinct from the similar Greek bird *Sylvia pallida*. Tristram could see no differences between the two and it is now generally accepted that they are conspecific and included in the genus *Hippolais*. Tristram's convict companion proved to be very knowledgeable and pointed out that the Olivaceous Warbler of the hillsides was somewhat darker and larger than that of the marshes and produced larger eggs. Tristram left the marsh and went some three miles into the hills, finding no trace of a larger warbler. On the way back he "shot *Sylvia olivetorum* (Olive Tree Warbler) and thus solved the mystery of the large Pallid Warbler." He saw several birds and later obtained a nest which was much inferior in neatness to that of the Pallid Warbler though the eggs were very similar but slightly larger. Back at the camp several nests and eggs awaited him but only that of a second Cetti's Warbler is mentioned in the record of the day's collecting.

Next day, going out onto the lake itself, many Black-headed and Mediterranean Gulls were encountered, but no breeding birds. Several species of terns were shot including Whiskered, Black and Lesser-crested Terns. On the open water Crested Coot, Wigeon and Pochard were the commonest birds. Tristram found the eggs of the Crested Coot to be slightly larger than those of the Common Coot which was absent from Halloula. Pushing through the reeds in the boat, several nests of Great Reed Warbler were found but these, though of similar general design to that of the Reed Warbler in England, were not of such careful construction. Savi's Warblers could be heard all around but their nests evaded the searchers. However, Tristram found some consolation in discovering the nest of the Aquatic Warbler and he "obtained the bird". French naturalists considered this species to be rare in Algeria but Tristram eventually "found it in small numbers in all suitable localities". The nest he found at Halloula was not suspended like that of the English Reed Warbler but was entwined with five or six reeds about three feet above the water; he usually found nests of this species to be supported on a tuft.

Water Rail and Moorhen were common breeders at Halloula but Tristram found only one nest of the Purple Gallinule; it was built very like that of the Coot and the eggs had larger spots and blotches than those of the Moorhen. For Little Grebe and Great Crested Grebe it was now late in the season and several pairs had young. "Still, fifty eggs of one and about a dozen of the other was not a bad morning's take." Now we would find this unacceptable but Tristram was, to some extent, living off the land as probably were the convicts. In marked contrast to these two species the Black-necked Grebe nests in colonies and one of such was found on the lake "more crowded than any rookery". There appeared to be a great consistency in the date of laying in this colony since in almost fifty completed clutches Tristram found all the eggs to be unincubated. He shot several birds but was puzzled by their sudden disappearance. It appeared that water-tortoises were carrying off the bodies as they fell in the water! No nests of duck were to be found as for them the breeding season had not yet begun. Similarly, there was no trace of Herons' nests though vast flocks came in to roost in the evening.

Tristram stayed late into the evening to watch the Herons coming in to the roost, first the Squacco Herons and then Buff-backed Herons (Cattle Egrets) and Night Herons with a straggling of Glossy Ibis (Plate 20). Back in camp the convicts had provided for the collection of yet more trophies "which consisted chiefly of *S. hippolais*, two of *S. elaica*, and one of *S. cetti*. It is not entirely clear what these were, birds or eggs, nor is it clear to what bird the first of these names applies. *S. hippolais* is presumably the Icterine Warbler, but this does not nest in Algeria so this would make the trophies birds rather than eggs. The other two names are Olivaceous Warbler and Cetti's Warbler; unfortunately Tristram's application of scientific names is somewhat inconsistent. Whilst the trophies were few, the news was better—that of a nest of the Golden Eagle and a "digging" of *Merops apiaster* (European Bee-eater) in a bank hard by. The colony of several holes gave up one egg whereas the eagle's nest contained two young. When he returned to camp, two nests of Fantail Warblers had been brought in by soldiers who had been cutting forage. Apparently the nests are usually found only by chance as they are often well hidden about a foot above ground level and are constructed "by entwining the living stems of grass with very fine cotton and spiders' webs". The eggs are laid and incubation begun before the nest is complete and as the male continues to build the walls of the nest higher and higher.

Tristram returned to Algiers for a time, following the same route along which he had first approached Lake Halloula, retracing it in June and reaching the lake again on 10 July, with the objective of observing the breeding ducks and herons. On this occasion he hired a professional chasseur and learned on reaching the lake that many eggs had been collected for him.

Unfortunately these were all sold to M. Verreaux's agent before Tristram's return. Tristram and his chasseur remained for two days at the lake and found two nests of White-headed Duck with clutches of eggs, one with three and one with eight. There were several pairs of Pochard and Red-crested Whistling Duck on the lake but no nests were found. The Ferruginous Duck was also abundant on the lake and one nest was discovered.

The heronries lay on the southern side of the lake where the water gave way to mud and the punt had to be abandoned. Beyond this point was a heronry of Squacco Herons, most of which had just begun incubation. The nests, composed of heaps of water weed and rushes, were supported on tufts of reeds and contained either three or four eggs.

Beyond this heronry was another of Cattle Egrets (Buff-backed Herons) where the nests were very closely packed, and amongst them were a few Glossy Ibis. Searching amongst the Egrets' nests revealed two Ibis nests which were easily identifiable because the colour of the eggs was a much deeper blue and much rounder and smoother. Apparently the French chasseurs had much reduced the Ibis numbers. Further back from the lake were a few nests of Black-crowned Night Herons. Tristram commented in his account of these birds that in Algeria, most unusually he never encountered one in the spotted first year plumage.

Next day Tristram's chasseur shot a Red-crested Whistling Duck from the nest which contained a single egg. The species bred sparingly at the lake and overwintered there. A surprise lay in store for the hunters as they stumbled upon a colony of Whiskered Terns which had taken over the vacated nests of the Black-necked Grebes. The young grebes were present in their hundreds, diving with their parents in the open lake. The terns on the lake fed extensively on large hairy caterpillars which abounded on the marshes at this time of the year but also took newts and frogs by plunging into the water.

The return across the lake was not without incident, for Tristram and his chasseur grounded their punt, depositing all the loosely packed boxes into the mud. Several were lost, which may account for some of the birds collected not being listed in the future Catalogue.

A year later, rejoining Tristram's second expedition, with the heronries of the partially drained lake now almost deserted behind them, they moved south through the Pass of Chiffa, heading for Medea. Here they saw numerous Crag Martins, and buzzards, kites and falcons. Tristram collected none of these but he had taken a Black-shouldered Kite (Plate 21) in Algiers. South of Medea, where Boghar could be seen near the top of the highest mountain in the area (ca. 4,000 feet), the expedition encountered a limestone escarpment with steep cliffs and here Golden Eagles abounded. It was almost certainly here that Tristram took a female bird "near Medeah, Algeria, 1856". The following day the party encounterd the bodies of two recently shot birds "a magnificent Lammergeyer and an Eagle Owl". The former is included in Tristram's Catalogue as a female taken at Boghar, Algeria, November 1856, but there is no record of the latter.

Boghar stands at the northern frontier of the Sahara and here the party encountered "an immense flock of Dotterel" and shot several for food ("tender and good"), only one female being reported in the Catalogue: Bou Guizoon, Sahara, 1-11-57 (56). Camp was made at night by a small salt lake which Tristram had visited in the previous spring. Thousands of Greater Flamingos were present which flew off in a rosy cloud and were too difficult to approach in order to get a shot. Thousands of birds, however, remained, and the White-winged Black Terns came in very close to their watchers. Many Black-winged Stilts nested along the edge of the lake and in the reeds were Waterhens and Allen's Gallinules. In the thickets Great Reed Warblers, Thrush Nightingales and Aquatic Warblers sang. Collared Pratincoles (Plate 22) nested in abundance, often laying their three eggs in the dried footprints of camels, and Kentish Plovers scurried along the water's edge. Tristram wrote that the three days spent at Bouguizoun "went but little way to exhaust the ornithological marvels . . . though I worked and noted from sunrise to sunset."

At Bouguizoon, Tristram was the guest of Sheik Bou Disah who had brought with him his Lanner Falcons (Plate 23). Each falconer had an assistant and was in charge of three falcons. On the march the falconer carried one bird on his wrist, one on his shoulder and a third on his head. The main prey, then as now, is the Houbara Bustard, but falcons were also flown at eagles, kites and sandgrouse, and, in the case of the larger Saker Falcon, at gazelle. For Tristram's entertainment a falcon was to be flown at a bustard. Having sighted the prey the falcon was transferred to the hand of the Sheik who unhooded it. A second falcon was readied and the first launched. A skilled bird rises well above the bustard and waits there. It is considered very bad form to cause the bustard to take wing before the falcons are ready. The first falcon is then trained to dive low over the prey as the second falcon arrives on the scene, so keeping the bustard on the ground and causing it to break into a run. Any attempt to rise from the ground elicits a dive on it by one of the falcons and the prey is thus chased until exhausted. "It is considered the excellency of a falcon to make these feints at the quarry till he is nearly exhausted, when the fatal swoop is made, and the bird instantly drops, struck dead by the hind claw [of the falcon] piercing its vertebrae." On the occasion of Tristram's entertainment three houbaras and some sandgrouse were captured by three falcons "and the chase was terminated merely on account of the fatigue of the horses."

Sandgrouse provide a different form of entertainment which involves a Raven. As soon as the sandgrouse (usually in a covey) rise from the ground the falcon is unhooded and flown, followed by two others, and the birds select three prey which they take in flight. With a good start, usually because the falcons are not released sufficiently quickly, the sandgrouse will circle upwards in an attempt to prevent the falcons getting above them, then scattering, often successfully escaping. The Raven, circling above the group, is there to attract the falcons higher, and is trained to circle upwards.

Gazelle hunting with a Saker Falcon (Plate 24) is much more dangerous for the bird which is often impaled on the horns of the gazelle and "it is not unusual for both pursuer and victim to fall dead at one mutual stroke." There was to be no gazelle hunting for Tristram. Had this taken place he may well have obtained a dead Saker for his collection, but as it was he offered 200 dollars for one in vain, discovering that each was regarded as having the same value as a thoroughbred horse. The Lanner Falcon he found was priced equally to the Saker so it was no wonder that he failed to obtain one on this occasion. Tristram wrote: "Indeed it would have been a crime of the blackest dye to have shot one had I had the opportunity." It therefore follows that the male Lanner recorded in his Catalogue as taken in "Tunis, 1856" was acquired by other means, or perhaps the same rules do not apply in Tunis!

As the expedition moved further south into the Sahara it is reasonable to assume that Tristram's infrequent mention of birds indicates few worthwhile observations. A few bustards, sandgrouse, ravens and vultures were mentioned before passing between the two Zaharez, shallow lakes, almost dry in summer, which are fed by mountain streams. The westernmost was, in Tristram's time, about thirty miles (48 km) long, the one to the east some twenty-five miles (40 km) long, in winter, and each some ten miles wide. In the November of 1856 they were only half this size. Flamingos were present in myriads but there were few other birds. Approaching Djelfa the party passed through a gorge of rock salt, disturbing numbers of Rock Doves from holes in which they were roosting. Djelfa itself is described as cold and bleak, being approximately 3,400 feet (1,370 metres) above sea level. Flocks of Crossbills combined with the temperature, which was low enough to show hoar frost in the early morning, reminded Tristram of Norway. Further into the desert was equally bleak with only a few Ravens and a single pair of Golden Eagles attracting any attention from the ornithologists.

After camping overnight at the caravanserai of Ain el Ibel, the plain ahead contained "an abundance of bustard, sandgrouse, jerboas and other small rodents". En route Tristram had shot three Wheatears, two males and a female, of what he considered to be a new species which he initially referred to as a Bush Chat *Saxicola Philothamna,* describing it as an inhabitant of the dayats (depressions which contain water in wet periods) and plains, and not of the ravines. Tristram (1859) published this as a new species and it became known for a time as Tristram's Chat. However, Dresser (1873) sank this as a synonym of the Red-rumped Chat *Saxicola moesta* Lichtenstein, 1823, under which name it appears in Tristram's Catalogue (1889). Nowadays it is referred to as the Buff-rumped Wheatear *Oenanthe moesta* but retains the alternative English name of Tristram's Wheatear (del Hoyo et al., 1979 to 2013) (Table 5). The collection of these wheatears was dated as 8 November 1856, at Ain el Ibel.

After arriving in Laghouat the expedition rested for a fortnight and Tristram explored the several oases which were to be found close to the town. The palm-groves contained thousands of migratory birds, many of which would have been familiar to an English ornithologist. Chiffchaffs, Willow Warblers and Whitethroats were to be found in all the gardens, and Swallows and House Martins abounded around the houses. Hoopoe and Southern Grey Shrike were recorded and Tristram (1860b) comments that "a random shot will startle from under the dates a dozen 'booma' or Little Owl." The dayats varied in size from 200 yards (183 metres) across to over a mile (1.6 km) in length; they attracted flocks of sandgrouse "and many desert larks of various little known species". In the evening Ravens returned to their roosts in flocks, much like English Rooks. On one occasion Tristram saw that one group of Ravens was chasing another species of bird which was obviously a falcon. He promptly shot it and it turned out to be a Barbary Falcon (Plate 25), a rare species usually associated with mountainous regions. Golden Eagles and Royal Kites were also seen "to carry on a peripheral and bloodless warfare with the Raven". There can be little doubt that many birds that Tristram saw went unidentified, particularly some of the warblers; others, such as the shrikes, were easily identifiable as there was only one species. This was the Southern Grey Shrike, referred to by Tristram as the Pallid Shrike, and it was abundant. A permanent resident of the dayats and oases it fed largely on scarabs and sometimes even on small birds.

During their stay in Laghouat, Tristram and Peed were entertained on several occasions by the commandant in a hall carpeted by skins of lions, leopards and ostriches which the commandant himself had shot. There they clearly lived off the land being provided with "courses consisting of gazelle, bustard, wild duck, wild boar and starling pasty, winding up with huge bunches of dates and wooden bowls of kouskousou, served *a l'Arabe*, all of us eating from the same dish." After leaving Laghouat, at the midday stop Tristram obtained a new species of Lark, which he considered to be similar to the Horned Lark of Sweden. In his Catalogue (1889) he refers to it as *Otocoris bilopha* (female, El Aghouat, 26-11-56), but as with many other scientific names used by Tristram, changes necessary to abide by the rules of zoological nomenclature have resulted in several Generic names being used; it is now named Temminck's (Horned) Lark *Eremophila bilopha*.

The following day Tristram missed a Golden Eagle at fifty yards and the eagle flew off. To soften the blow, Tristram immediately spotted an Arabian Kite, probably a Black Kite, in the same dayat. He also watched a group of birds which he named "Sociable Desert Thrush *(Malurus Numidicus)* as they ran up the bushes in the fashion of woodpeckers, often assisting themselves by the beak ... and descend again at the other side to halt again at the next shrub or tree." This species is now known as the Fulvous Babbler *Turdoides fulvus* (del Hoyo et al., 1988).

According to Tristram "the capture of an Ostrich is the greatest feat of hunting to which a Saharan sportsman aspires" and a hopeless pursuit! He therefore declined the invitation from the local kadi to take part in such a hunt, instead remaining in the dayat in which they had stopped in order to shoot Stock Doves *Columba oenas* for dinner. The next day dinner consisted of Barbary Partridge, a covey of which sprang from brushwood, the guns being quick enough to secure a few.

Berryan was the next town on the route south and Tristram found the groves and gardens far superior to those in Laghouat. Eighteen brace of doves were shot, presumably for feeding purposes, and numerous warblers from Europe were about, together with many Ravens and a few Kestrels. Sometime later, Peed shot a single Dupont's Lark (male, Waregla, Sahara, December 1856) which was very rare and only found in the ravines of the Wed N'ca. Little of ornithological interest was recorded in passing through Melika and on to Ghardaia. Turning east the next objective was Waregla, the last oasis of North Africa, with no other break until the Mountains of the Moon and Timbuctoo. The way forward was desolate and two days before arriving in Waregla, as they headed south, only two living things were seen throughout the day—"a curious white scorpion and a Desert Lark which must have lost its way."

In Waregla (Ouargla), 32 degrees north of the equator, Tristram shot three Little Owls, and having left the town, and approaching N'goussa, came across many Cream-coloured Coursers "but, loaded with ball, had no opportunity of obtaining this rare bird". Two days before Christmas, 1856, the cavalcade arrived at the small lake of Ain Bahrdad, which swarmed with birds, and collected a pair of Little Stints (Plate 26). Tristram obtained what he considered to be a new species of Fantail Warbler which eventually turned out to be the Scrub Warbler. Here in the south he did find a new Warbler which became Tristram's Warbler *Sylvia deserticola* (Plate 27). At Blad el Amer a few snipe and Sandpipers were seen feeding in a fetid ditch as the party passed on to Tuggert. Here an Arab officer introduced himself to Tristram and invited him to see his falcons—"a mews of seven well-trained birds of the scarcely known sakkr falcon (*Falco sacer*)", now *Falco cherrug*. At this time the species was found only in the southern spurs of the Atlas mountains, between Biskra and Boucada.

News had obviously spread of Tristram's arrival in Tuggert as he was visited by a fellow Northumbrian whose family was known to him. The visitor, who was awaiting his commission in the Foreign Legion, took Tristram out of the town to a wide open salt lake where duck, snipe and plovers abounded. There were many flocks of Tufted Duck and Wigeon but amongst them rarer English duck such as White-eyed and Red-crested Pochard. The party left Tuggert almost immediately because of lack of available food for the horses, and in the village of Zouia Tristram attempted, unsuccessfully, to collect specimens of Great White Egret, Little Egret and Purple Heron (Plate 28), all of which evaded him, though he shot a few duck for dinner. Further north-east at the well of Elouibed Tristram "had the good fortune to procure three specimens of a lark new to science, and which I never discovered elsewhere." He called this bird the Sand Lark *Galerida arenicola* but it is now considered as a sub-species of the Crested Lark *Galerida cristata arenicola*.

Having stayed three days in the oasis of Souf, at El Oed, but finding little of ornithological interest, the expedition proceeded north-east towards the Tunisian border, then turned north towards Biskra. Arriving in the town, Tristram found "lovely open fields" in place of the expected masses of palms and the Wed Biskra, "a scanty stream in a wide gravelly bed". "Ever and anon the snipe or the green sandpiper, and sometimes the wild duck, rise from behind a palm . . . while the Turtle Dove . . . steals from the far side of the tree." From Biskra to the oasis of El Kantara, the most northerly limit of the palm tree and from there into the ridges and gorges of the Atlas, they journeyed to where the expedition found both circling vultures and French roads. Tristram saw the Saker Falcon which he longed to collect and which evaded him. Here Tristram's record came to a most abrupt end, but with a sighting that would encourage a return to the Regency of Tunis.

This account of Tristram's birds, taken mainly from his *Great Sahara* (1860b) covers only a small part of his activities in the Great Desert. He was shot at by his guide who also stole from him, thrown from his horse on more than one occasion, engulfed in sandstorms and ambushed by tribesmen. He was also wined and dined royally on many occasions when he reached what passed for civilization and if his account is to be believed, which it surely is, then he took all this with a simple equanimity which is shown by few. His medical advisors back in England would probably have been taken aback by his idea of convalescence from illness, but it certainly seemed to have worked.

## Table 5: Birds observed by Tristram in North Africa

| Current Nomenclature | | Tristram's Nomenclature | |
|---|---|---|---|
| Common Ostrich | *Struthio camelus* | Ostrich | *Struthio Camelus* |
| Little Grebe | *Tachybaptus ruficollis* | Little Grebe | *Podiceps minor* |
| Great-crested Grebe | *Podiceps cristatus* | Great-crested Grebe | *Podiceps Cristatus* |
| Black-necked Grebe | *Podiceps nigricollis* | Black-necked Grebe | ***Podiceps nigricollis*** |
| Black-necked Grebe | *Podiceps nigricollis* | (Eared Grebe) | *Podiceps Auritus* |
| Cory's Shearwater | *Calonectris diomadea* | Mediterranean Shearwater | ***Puffinus kuhlu*** |
| European Storm Petrel | *Procellaria pelagica* | Storm Petrel | ***Procellaria pelagica*** |
| Pygmy Cormorant | *Phalacrocorax pygmaeus* | Pygmy Cormorant | ***Phalacrocorax pygmaeus*** |
| White Stork | *Ciconia ciconia* | White Stork | *Ciconia Alba* |
| Common Heron | *Ardea cinerea* | Common Heron | *Ardea Cinerea* |
| Purple Heron | *Ardea purpurea* | Purple Heron | *Ardea Purpurea* |
| Great Egret | *Ardea alba* | White Egret | *Herodias Alba* |
| Little Egret | *Egretta garzetta* | Lesser Egret | *Herodias Garzetta* |
| Cattle Egret | *Bubulcus ibis* | Buff-backed Heron | ***Bubulcus Ibis*** |
| Squacco Heron | *Ardeola ralloides* | Squacco Heron | ***Buphus Comatus*** |
| Little Bittern | *Ixobrychus minutus* | Little Bittern | ***Ardetta Minuta*** |
| Bittern | *Botaurus stellaris* | Bittern | ***Botaurus Stellaris*** |
| Night Heron | *Nycticorax nycticorax* | Night Heron | ***Nycticorax Griseus*** |
| Greater Flamingo | *Phoenicopterus ruber* | Flamingo | *Phoenicopterus Antiquorum* |
| Northern Bald Ibis | *Geronticus eremita* | Bald Ibis | *Geronticus Comatus* |
| Glossy Ibis | *Plegadis falcinellus* | Glossy Ibis | *Falcinellus Igneus* |
| Bean Goose | *Anser fabilis* | Bean Goose | *Anser Segetum* |
| Ruddy Shelduck | *Tadorna ferruginea* | Ruddy Shieldrake | ***Casarla Rutila*** |
| Common Shelduck | *Tadorna tadorna* | Common Shieldrake | *Tadorna Vulpanser* |
| Mallard | *Anas platyrhynchos* | Wild Duck | *Anas platyrhynchos* |

| Current Nomenclature | | Tristram's Nomenclature | |
|---|---|---|---|
| Gadwall | *Anas strepera* | Gadwall | ***Anas Strepera*** |
| Northern Shoveler | *Anas clypeata* | Shoveller | ***Rhynchaspis Clypeata*** |
| Gargany | *Anas querquedula* | Gargany | ***Querquedula circia*** |
| Eurasian Teal | *Anas crecca* | Teal | *Querquedula Crecca* |
| Northern Pintail | *Anas acuta* | Pintail | *Dafila Acuta* |
| Eurasian Wigeon | *Anas penelope* | Wigeon | *Mareca Penelope* |
| Tufted Duck | *Aythya fuligula* | Tufted Duck | ***Fuligula Cristata*** |
| Common Pochard | *Aythya ferina* | Pochard | *Fuligula Ferina* |
| Ferruginous Duck | *Aythya nyroca* | White-eyed Duck | ***Nyroca Leucophthalma*** |
| Red-crested Pochard | *Netta rufina* | Red-crested Whistling Duck | ***Calichen Rufina*** |
| White-headed Duck | *Oxyura leucocephala* | White-headed Duck | ***Erismatura Mersa*** |
| Griffon Vulture | *Gyps fulvus* | Griffon Vulture | ***Gyps Fulvus*** |
| Lappet-faced Vulture | *Torgos tracheliotus* | Nubian Vulture | *Otogyps Nubicus* |
| Egyptian Vulture | *Neophron percnopterus* | Egyptian Vulture | *Neophron Percnopterus* |
| Lammergeyer | *Gypaetus barbatus* | Lammergeyer | ***Gypaetus Barbatus*** |
| Golden Eagle | *Aquila chrysaetos* | Golden Eagle | ***Aquila Chrysaetos*** |
| Montagu's Harrier | *Circus pygargus* | Montagu's Harrier | ***Circus pygargus*** |
| Short-toed Eagle | *Circaetus gallicus* | Short-toed Eagle | *Circaetus Gallicus* |
| Sparrowhawk | *Accipiter nisus* | Sparrowhawk | ***Accipiter nisus*** |
| Long-legged Buzzard | *Buteo rufinus* | Long-legged Buzzard | ***Buteo desertorum*** |
| Bonelli's Eagle | *Hieraaetus fasciatus* | Bonelli's Eagle | ***Nsaetus fasciatus*** |
| Black-shouldered Kite | *Elanus caeruleus* | Black-shouldered Kite | ***Elanus caerulcus*** |
| Honey Buzzard | *Pernis apivorus* | Honey Buzzard | ***Pernis apivorus*** |
| Saker Falcon | *Falco cherug* | Saker Falcon | *Falco Sacer* |
| Lanner Falcon | *Falco biarmicus* | Lanner Falcon | ***Falco Lanarius*** |
| Barbary Falcon | *Falco peregrinoides* | Barbary Falcon | ***Falco Barbarus*** |
| Hobby | *Falco subbuteo* | Hobby | *Hypotriorchis Subbuteo* |

| Current Nomenclature | | Tristram's Nomenclature | |
|---|---|---|---|
| Kestrel | *Falco tinnunculus* | Kestrel | ***Tinnunculus Alaudarius*** |
| Kite | *Milvus milvus* | Kite | *Milvus Regalis* |
| Black Kite | *Milvus migrans* | Egyptian Kite | *Milvus Aegyptus* |
| Barbary Partridge | *Alectoris barbara* | Barbary Partridge | ***Caccabis Petrosa*** |
| Common Quail | *Coturnix coturnix* | Quail | ***Coturnix Communis*** |
| Common Crane | *Grus grus* | Crane | *Grus Cinerea* |
| Demoiselle Crane | *Anthropoides virgo* | Numidian Crane | ***Anthropoides Virgo*** |
| Water Rail | *Rallus aquaticus* | Water Rail | *Rallus Aquaticus* |
| Baillon's Crake | *Portzana pusilla* | Baillon's Crake | *Gallinula Bailloni* |
| Purple Gallinule | *Porphyrio porphyrio* | Great Purple Gallinule | ***Porphyrio Hyacinthus*** |
| Eurasian Coot | *Fulica atra* | Coot | ***Fulica Atra*** |
| Crested Coot | *Fulica cristata* | Crested Coot | ***Fulica cristata*** |
| Little Bustard | *Tetrax tetrax* | Little Bustard | ***Otis Tetrax*** |
| Houbara Bustard | *Chlamydotis undulata* | Ruffed Bustard | ***Houbara Undulata*** |
| Black-winged Stilt | *Himantopus himantopus* | Black-winged Stilt | ***Himantopus Melanopterus*** |
| Avocet | *Recurvirostra avocetta* | Avocet | *Recurvirostra Avocetta* |
| Stone Curlew | *Burhinus oedicnemus* | Norfolk Plover | ***Oedicnemus Crepitans*** |
| Cream-coloured Courser | *Cursorius cursor* | Cream-coloured Courser | ***Cursorius Gallicus*** |
| Collared Pratincole | *Glareola pratincola* | Collared Pratincole | ***Glareola Pratincola*** |
| Northern Lapwing | *Vanellus vanellus* | Lapwing | ***Vanellus Cristatus*** |
| Spur-winged Lapwing | *Vanellus spinosus* | Spur-winged Plover | ***Haplopterus spinus*** |
| Ringed Plover | *Charadrius hiaticuls* | Ringed Plover | ***Aegialites hiaticula*** |
| Little Ringed Plover | *Charadrius dubius* | Little Ringed Plover | *Aegialites Minor* |
| Kentish Plover | *Charadrius alexandrinus* | Kentish Plover | ***Aegialites Cantianus*** |
| Dotterell | *Eudromis morinellus* | Dotterell | ***Eudromius Morinellus*** |

| Current Nomenclature | | Tristram's Nomenclature | |
|---|---|---|---|
| Slender-billed Curlew | *Numenius tenuirostris* | Slender-billed Curlew | *Numenius Tenuirostris* |
| Common Redshank | *Tringa totanus* | Redshank | **Gambetta Calidris** |
| Green Sandpiper | *Tringa ochropus* | Green Sandpiper | *Totanus Ochropus* |
| Wood Sandpiper | *Tringa glareola* | Wood Sandpiper | *Totanus Glareola* |
| Common Sandpiper | *Actitis hypoleucos* | Common Sandpiper | *Totanus Hypoleucos* |
| Temminck's Stint | *Calidris temminckii* | Temminck's Stint | *Tringa Temminkii* |
| Dunlin | *Calidris alpina* | Dunlin | ***Tringa Alpina*** |
| Curlew Sandpiper | *Calidris ferruginea* | Curlew Sandpiper | ***Tringa subarquata*** |
| Common Snipe | *Gallinago gallinago* | Common Snipe | ***Gallinago media*** |
| Jack Snipe | *Lymnocryptes minimus* | Jack Snipe | *Gallinago Gallinula* |
| Ruff | *Philomachus pugnax* | Ruff | ***Machetes Pugnax*** |
| Whiskered Tern | *Chlidonias hybridus* | Whiskered Tern | ***Hydrochelidon Hybrida*** |
| Gull-billed Tern | *Gelochelidon nilotica* | Gull-billed Tern | *Gelochelidon Anglica* |
| Little Tern | *Sterna albifrons* | Least Tern | *Sterna Minuta* |
| White-winged Black Tern | *Chlidonias leucopterus* | White-winged Black Tern | *Hydrochelidon Leucoptera* |
| Black Tern | *Chlidonias niger* | Black Tern | ***Hydrochelidon Nigra*** |
| Lesser Crested Tern | *Thalasseus bengalensis* | Lesser Crested Tern | *Sterna media* |
| Slender-billed Gull | *Larus genei* | Slender-billed Gull | ***Larus gelastes*** |
| Little Gull | *Larus minutus* | Little Gull | ***Larus minutus*** |
| Pin-tailed Sand Grouse | *Pterocles alchata* | Pin-tailed Sand Grouse | ***Pterocles Alchata*** |
| Spotted Sand Grouse | *Pterocles senegalus* | Senegal Sand-Grouse | *Pterocles Senegalus* |
| Black-bellied Sandgrouse | *Pterocles orientalis* | Black-bellied Sandgrouse | *Pterocles orientalis* |
| Crowned Sandgrouse | *Pterocles coronatus* | Spotted Sand Grouse | ***Pterocles Coronatus*** |
| Wood Pigeon | *Columba palumbus* | Wood Pigeon | ***Columba palumbus*** |

| Current Nomenclature | | Tristram's Nomenclature | |
|---|---|---|---|
| Rock Dove | *Columba livia* | Rock Dove | ***Columba Livia*** |
| Stock Dove | *Columba oenas* | Stock Dove | ***Columba Oenas*** |
| Turtle Dove | *Streptopelia turtur* | Turtle Dove | ***Turtur communis*** |
| Laughing Dove | *Streptopelia senegalensis* | Egyptian Turtle Dove | ***Turtur Aegyptiacus*** |
| Cuckoo | *Cuculus canorus* | Cuckoo | ***Cuculus canorus*** |
| Great Spotted Cuckoo | *Cuculus glandarius* | Great Spotted Cockoo | ***Coccystes glandarius*** |
| Tawny Owl | *Strix aluco* | Tawny Owl | ***Syrnium Numidia*** |
| European Scops Owl | *Otus scops* | Scops Owl | ***Scops Zorva*** |
| Eagle Owl | *Bubo bubo* | Egyptian Great Owl | ***Ascalaphia Savignii*** |
| Little Owl | *Athene noctua* | Algerian Little Owl | ***Athene Glaux*** |
| Alpine Swift | *Tachymarptis melba* | White-bellied Swift | ***Cypselus Melba*** |
| Common Swift | *Apus apus* | Common Swift | ***Cypselus Apus*** |
| White-breasted Kingfisher | *Halcyon smyrnensis* | White-breasted Kingfisher | ***Halcyon smyrnensis*** |
| Bee-eater | *Merops apiaster* | Bee-eater | ***Merops Apiaster*** |
| European Roller | *Coracias garrulus* | Roller | ***Coracias garrulus*** |
| Hoopoe | *Upupa epops* | Hoopoe | ***Upupa Epops*** |
| Eurasian Wryneck | *Jynx torquilla* | Wryneck | ***Jynx torquilla*** |
| Great Spotted Woodpecker | *Dendrocopus major* | Algerian Pied Woodpecker | ***Picus numidicus*** |
| Green Woodpecker | *Picus viridis* | Green Woodpecker | ***Cecinus algirus*** |
| Crested Lark | *Galerida cristata* | Crested Lark | ***Galerida Cristata*** |
| Temminck's Horned Lark | *Eremophila bilopha* | Desert Horned Lark | ***Otocorys Bilopha*** |
| Woodlark | *Lullula arborea* | Woodlark | ***Alauda arborea*** |
| Greater Hoopoe Lark | *Alaemon alaudipes* | Bifasciated Lark | ***Certhilauda Desertorum*** |
| Dupont's Lark | *Cherphosilus duponti* | Dupont's Lark | ***Certhilauda Duponti*** |
| Bar-tailed Lark | *Ammomanes cinctura* | Bar-tailed Lark | ***Ammomanes cincturus*** |
| Desert Lark | *Ammomanes deserti* | Desert Lark | ***Ammomanes Isabellina*** |

| Current Nomenclature | | Tristram's Nomenclature | |
| --- | --- | --- | --- |
| Thick-billed Lark | *Rhamphocoris clotbey* | Cavaignac's Lark | ***Rhamphocoris Clot-bey*** |
| Bimaculated Lark | *Melanocorypha bimaculata* | Bimaculated Lark | ***Melanocorypha bimaculata*** |
| Calandra Lark | *Melanocorypha calandra* | Calandra Lark | ***Melanocorypha Calandra*** |
| Greater Short-toed Lark | *Calandrella brachydactyla* | Short-toed Lark | *Calandrella Brachydactyla* |
| Greater Short-toed Lark | *Calandrella brachydactyla* | Reboud's Lark | *Calandrella Reboudia* |
| Lesser Short-toed Lark | *Calandrella rufescens* | Lesser Short-toed Lark | ***Calandrella minor*** |
| Horned Lark | *Eremophila alpestris* | Desert Horned Lark | *Otocoys Bilopha* |
| Grested Lark | *Galerida cristata* | Crested Lark | ***Galerida Cristata*** |
| Crested Lark | *Galerida cristata macrorhyncha* | Long-billed Crested Lark | ***Galerida Macrorhyncha*** |
| Crested Lark | *Galerida cristata arenicola* | Sand Lark | ***Galerida Arenicola*** |
| Crested Lark | *Galerida cristata* | (Pale Desert Lark) | *Ammomanes Pallida* |
| Crested Lark | *Galerida cristata* | (Little Desert Lark) | *Ammomanes Regulus* |
| Grested Lark | *Galerida cristata* | (Abyssinian Lark) | ***Galerida Abyssinica*** |
| Grested Lark | *Galerida cristata* | (Isabelline Crested Lark) | ***Galerida Isabellina*** |
| Crested Lark | *Galerida cristata* | (Salvin's Lark) | *Certhilauda Salvini* |
| Barn Swallow | *Hirundo rustica* | Chimney Swallow | *Hirundo Rustica* |
| Red-rumped Swallow | *Hirundo daurica* | Red-rumped Swallow | ***Hirundo rufula*** |
| European Crag Martin | *Hirundo rupestris* | Rock Swallow | ***Cotyle Rupestris*** |
| Common Sand Martin | *Riperia riperia* | Sand Martin | *Cotyle Riperia* |
| Common House Martin | *Delichon urbica* | House Martin | *Chelidon Urbica* |
| White Wagtail | *Motacilla alba* | White Wagtail | ***Motacilla Alba*** |
| Yellow Wagtail | *Motacilla flava* | Yellow Wagtail | ***Budytes Flava*** |

| Current Nomenclature | | Tristram's Nomenclature | |
|---|---|---|---|
| Yellow Wagtail | *Motacilla flava* | (Ashy-headed Wagtail) | ***Motacilla cinereocapilla*** |
| Tawny Pipit | *Anthus campestris* | Tawny Pipit | ***Anthus Campestris*** |
| Meadow Pipit | *Anthus pratensis* | Meadow Pipit | ***Anthus Pratensis*** |
| Eurasian Tree Pipit | *Anthus trivialis* | Tree Pipit | ***Anthus Arboreus*** |
| Water Pipit | *Anthus spinoletta* | Water Pipit | ***Anthus spinoletta*** |
| Common Bulbul | *Pycnonotus barbatus* | Garden Bulbul | ***Pycnonotus barbatus*** |
| White-eyed Bulbul | *Pycnonotus xanthopygus* | White-eyed Bulbul | ***Pycnonotus xanthopygus*** |
| Blue Rock Thrush | *Monticola solitarius* | Blue Thrush | ***Monticola Cyanea*** |
| Rufous-tailed Rock Thrush | *Monticola saxatilis* | Rock Thrush | ***Monticola Saxatilis*** |
| Black Wheatear | *Oenanthe leucura* | Black Wheatear | ***Dromolaea Leucura*** |
| White-crowned Wheatear | *Oenanthe leucopyga* | White-rumped Rock Chat | ***Dromolaea Leucopygia*** |
| White-crowned Wheatear | *Oenanthe leucopyga* | (White-headed Rock Chat) | *Dromolaea Leucocephala* |
| Buff-rumped Wheatear | *Oenanthe moesta* | Bush Chat | ***Saxicola Philothamma*** |
| Northern Wheatear | *Oenanthe oenanthe* | Wheatear | ***Saxicola Oenanthe*** |
| Black-eared Wheatear | *Oenanthe hispanica* | Stapazine Chat | ***Saxicola Stapazina*** |
| Black-eared Wheatear | *Oenanthe hispanica* | (Eared Chat) | *Saxicola Aurita* |
| Mourning Wheatear | *Oenanthe lugens* | Mourning Chat | ***Saxicola Lugens*** |
| Mourning Wheatear | *Oenanthe lugens halophila* | Salt-loving Chat | *Saxicola Halophila* |
| Desert Wheatear | *Oenanthe deserti homochroa* | Solitary Chat | *Saxicola Homochroa* |
| Mourning Wheatear | *Oenanthe lugens persica* | Mourning Wheatear | ***Saxicola persica*** |
| Desert Wheatear | *Oenanthe deserti* | Desert Chat | ***Saxicola Deserti*** |
| Whinchat | *Saxicola rubetra* | Whinchat | ***Pratincola Rubetra*** |
| Stonechat | *Saxicola rubicola* | Stonechat | ***Pratincola Rubicola*** |

| Current Nomenclature | | Tristram's Nomenclature | |
|---|---|---|---|
| Moussier's Redstart | *Phoenicurus moussieri* | Moussier's Warbler | ***Ruticilla Moussieri*** |
| Common Redstart | *Phoenicurus phoenicurus* | Redstart | ***Ruticilla phoenicurus*** |
| Black Redstart | *Phoenicurus ochruros* | Tithys Redstart | ***Ruticilla tithys*** |
| Robin | *Erithacxus rubecula* | Robin | *Dendalus Rubecula* |
| Bluethroat | *Luscinia svecica* | Blue-throated Robin | ***Erithacus cyanecula*** |
| Thrush Nightingale | *Luscinia luscinia* | Thrush Nightingale | ***Erithacus luscinia*** |
| Blackcap | *Sylvia atricapilla* | Black Cap | ***Curruca Atricapilla*** |
| Garden Warbler | *Sylvia borin* | Garden Warbler | ***Curruca Hortensis*** |
| Orphaean Warbler | *Sylvia hortensis* | Orphaean Warbler | ***Curruca Orphea*** |
| Subalpine Warbler | *Sylvia subalpina* | Subalpine Warbler | ***Sylvia subalpina*** |
| Sardinian Warbler | *Sylvia melanocephala* | Sardinian Warbler | ***Sylvia melanocephala*** |
| Lesser Whitethroat | *Sylvia curruca* | Lesser Whitethroat | *Sylvia Curruca* |
| Whitethroat | *Sylvia communis* | Whitethroat | *Sylvia Cinerea* |
| Spectacled Warbler | *Sylvia conspicillata* | Spectacled Warbler | ***Sylvia Conspicillata*** |
| Tristram's Warbler | *Sylvia deserticola* | Desert Warbler | *Sylvia Deserticola* |
| Dartford Warbler | *Sylvia undata* | Dartford Warbler | ***Melizophilus Provincialis*** |
| Melodious Warbler | *Hippolais polyglotta* | Melodious Willow Wren | ***Hippolais Polyglotta*** |
| Olivaceous Warbler | *Hippolais pallida* | Pallid Warbler | *Hippolais Pallida* |
| Olive Tree Warbler | *Hippolais olivetorum* | Olive Tree Warbler | ***Hippolais olivetorum*** |
| Western Olive Tree Warbler | *Hippolais olivetorum* | (Western Olive Tree Warbler) | ***Hippolais opaca*** |
| Great Reed Warbler | *Acrocephalus arundinaceus* | Great Sedge Warbler | ***Calamoherpe Turdoides*** |
| Aquatic Warbler | *Acrocephalus paludicola* | Aquatic Warbler | ***Acrocephalus aquaticus*** |
| Sedge Warbler | *Acrocephalus schoenobaenus* | Sedge Warbler | ***Acrocephalus phragmites*** |
| Moustached Warbler | *Acrocephalus melanopogon* | Moustached Warbler | ***Lusciniola melanopogon*** |

| Current Nomenclature | | Tristram's Nomenclature | |
|---|---|---|---|
| Cetti's Warbler | *Cettia cetti* | Cetti's Warbler | ***Cettia Sericea*** |
| Fantail Warbler | *Cisticola juncidis* | Fantail Warbler | ***Cisticola cursitans*** |
| Savi's Warbler | *Locustella luscinoides* | Savis Warbler | ***Lusciniopsis Savii*** |
| Rufous Scrub Robin | *Cercotrichus galactotes* | Rufous Sedge Warbler | ***Aedon Galactodes*** |
| European Pied Flycatcher | *Muscicapa atricapilla* | Pied Flycatcher | ***Muscicapa atricapilla*** |
| Scrub Warbler | *Scotocerca inquieta* | Desert Fantail | ***Drymoica Straticeps*** |
| Graceful Prinia | *Prinia gracilis* | Graceful Prinia | ***Burnesia gracilis*** |
| Fulvous Babbler | *Turdoides fulvus* | Nubian Malurus | ***Crateropus Fulvus*** |
| Willow Warbler | *Phylloscopus trochilus* | Willow Wren | ***Phylloscopus Trochilus*** |
| Chiffchaff | *Phylloscopus collybita* | Chiffchaff | ***Phylloscopus Rufus*** |
| Western Bonelli's Warbler | *Phylloscopus bonelli* | Bonelli's Warbler | ***Phylloscopus Bonellii*** |
| North African Blue Titmouse | *Cyanistes caeruleus ultramarinus* | Teneriffe Blue Titmouse | ***Parus teneriffae*** |
| Southern Grey Shrike | *Lanius meridionalis* | Pallid Shrike | ***Lanius Dealbatus*** |
| Southern Grey Shrike | *Lanius meridionalis* | Pallid Shrike | ***Lanius algeriensis*** |
| Southern Grey Shrike | *Lanius meridionalis* | Pallid Shrike | ***Lanius auriculatus*** |
| Southern Grey Shrike | *Lanius meridionalis* | Pallid Shrike | ***Lanius elegans*** |
| Southern Grey Shrike | *Lanius meridionalis* | Pallid Shrike | ***Lanius hemileucurus*** |
| Masked Shrike | *Lanius nubicans* | Masked Shrike | ***Lanius nubicus*** |
| Black-crowned Bush Shrike | *Tchagra senegalus* | Black-crowned Bush Shrike | ***Telephonus cucculatus*** |
| Common Wren | *Troglodytes troglodytes* | Wren | ***Troglodytes parvulus*** |
| Wallcreeper | *Tichodroma muraria* | Wallcreeper | ***Tichodroma muraria*** |
| Palestine Sunbird | *Nectarinia oseae* | Palestine Sunbird | ***Cinnyris oseae*** |

| | Current Nomenclature | | Tristram's Nomenclature | |
|---|---|---|---|---|
| Eurasian Jay | *Garrulus glandarius* | Algerian Black-headed Jay | ***Garrulus cervicalis*** |
| Magpie | *Pica pica* | Algerian Magpie | ***Pica Mauritanica*** |
| Raven | *Corvus corax* | Raven | ***Corvus Corax*** |
| Jackdaw | *Corvus monedula* | Jackdaw | ***Corvus Monedula*** |
| Chough | *Pyrrhocorax pyrrhocorax* | Chough | ***Pyrrhocorax Graculus*** |
| Common Starling | *Sturnus vulgaris* | Common Starling | ***Sturnus vulgaris*** |
| Spotless Starling | *Sturnus unicolor* | Black Starling | ***Sturnus Unicolor*** |
| Spanish Sparrow | *Passer hispaniolensis* | Spanish Sparrow | ***Passer Salicarius*** |
| Rock Sparrow | *Pteronia petronia* | Rock Sparrow | ***Petronia stulta*** |
| Italian Sparrow | *Passer italiae* | Cisalpine Sparrow | ***Passer Italiae*** |
| Desert Sparrow | *Passer simplex* | Desert Sparrow | ***Corospiza Simplex*** |
| Chaffinch | *Fringilla coelebs africana* | Algerian Chaffinch | ***Fringilla spodiogenys*** |
| Brambling | *Fringilla montifringilla* | Brambling | ***Montifringilla nivalis*** |
| Hawfinch | *Coccothraustes coccothraustes* | Hawfinch | ***Coccothraustes vulgaris*** |
| Serin | *Serinus serinus* | Serin Finch | ***Serinus hortulanus*** |
| Greenfinch | *Carduelis chloris* | Algerian Greenfinch | ***Ligurinius aurantiventris*** |
| Trumpeter Finch | *Bucanetes githagineus* | Vinous Grosbeak | ***Pyrrhula Githaginea*** |
| Common Crossbill | *Loxia recurvirostra* | Crossbill | ***Loxia curvirostra*** |
| Corn Bunting | *Miliaria calandra* | Bunting | ***Emberiza miliaria*** |
| Cirl Bunting | *Emberiza cirlus* | Cirl Bunting | ***Emberiza Cirlus*** |
| Rock Bunting | *Emberiza cia* | Meadow Bunting | *Emberiza Cia* |
| Cretzschmar's Bunting | *Emberiza caesia* | Cretzschmar's Bunting | ***Emberiza cacsia*** |
| House Bunting | *Emberiza strioloata* | House Bunting | *Fringillaria Saharae* |

Bold type indicates birds collected by Tristram.

Plate 18: Alfred Newton, first Professor of Zoology and Comparative Anatomy in the University of Cambridge. He was a very close friend of Tristram and they consulted each other frequently; *Ibis* Jubilee Supplement 1908, Vol. 50:19. Reproduced with the permission of the British Ornithologists' Union, www.bou.org.uk.

Plate 19: Montagu's Harrier *Circus pygargus*, a bird of the plains and lowlands of North Africa and Palestine.

Plate 20: Glossy Ibis *Plegadis falcinellus*, a bird of marshland areas where it often bred in North Africa in colonies with herons and egrets.

Plate 21: Black-shouldered Kite *Elanus caeruleus* is found in areas of scattered trees and woodland edges and is mainly a resident bird in North Africa.

Plate 22: Collared Pratincole *Glareola pratincola* is to be found on dried-out mudflats and plains where it nests on the ground colonially, usually near some water.

Plate 23: Lanner Falcon *Falco biarmicus* is the common bird used in falconry in North Africa and is flown at birds.

Plate 24: Sakker Falcon *Falco cherrug*, larger than the Lanner Falcon and sometimes flown at small antelopes.

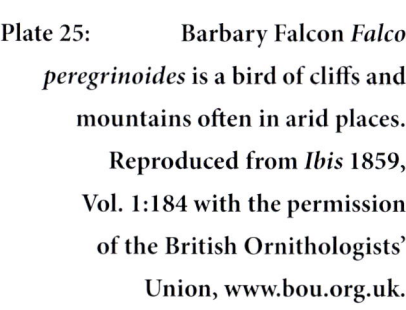

Plate 25: Barbary Falcon *Falco peregrinoides* is a bird of cliffs and mountains often in arid places. Reproduced from *Ibis* 1859, Vol. 1:184 with the permission of the British Ornithologists' Union, www.bou.org.uk.

Plate 26: Little Stint *Calidris minuta* (large flock) and Curlew Sandpiper *Calidris ferruginea* (a single, slightly larger bird in the centre of the image) are migratory birds of the shore or the edges of inland lakes.

Plate 27: Tristram's Warbler *Sylvia deserticola* is found in scrubland, often in evergreen oak or juniper on high ground. From Tristram's *Fauna and Flora of Palestine* 1884.

Plate 28: Purple Heron *Ardea purpurea* is mainly a summer visitor to North Africa and often breeds in marshy areas in colonies with other species of heron.

Plate 29: Red Kite *Milvus milvus* is a bird of woodland, and hunts over open country.

Plate 30: Tawny eagle *Aquila repax* is a species found in both mountainous areas and the plains, and is mainly resident where it breeds.

— 6 —

# The Sands of the Sahara: in Tunisia

GRIFFON-VULTURE. (*Gyps fulvus.*)

With the first part of the planned expedition completed, Peed returned to England and Tristram met up with Osbert Salvin and Wilfred Simpson who were to continue with him in their journeying through what was then the Regency of Tunis and into eastern Algeria. This part of the expedition was planned to take place from the latter part of February until July of 1857 and their base was set up initially in the city of Tunis. From here they made excursions to Carthage and Oudena in the immediate vicinity, to Sousa and el Djem in the south and Bizerta in the north. Both Tristram (1860b) and Salvin (1859) gave accounts of the expedition, the latter being more a listing of the birds rather than of observations. During this period Tristram collected few birds, his interests apparently lying elsewhere.

The first month was largely spent organizing and planning routes. Tristram had previously arranged for the horses and camping equipment used in the first half of the expedition to be sent to the most easterly station of the French occupation of Algeria, Souk Harras, to which they travelled from Tunis at the end of March, having been joined by Tristram. Whilst in Tunis Tristram collected a Greater Flamingo from the Lagoon and a Kentish Plover on the coast north of the town. Here they were partially influenced by the accounts of two Hungarian hunters whom they had met. It was decided that Simpson should seek suitable camping grounds and "examine the habits of the Raptorials at home", whilst Salvin should go south west to prospect the desert birds; Tristram was to regress to the Tunisian frontier and investigate the Cork forests and lakes to the north.

Leaving Kef (not to be confused with Kef Laks), with the intention of returning, some half a day's journey westward, a family party of two adult and four young Common Crossbill *Loxia curvirostra* was observed on 26 March in pines; the young were newly fledged. The expedition spent a single night with the lawless frontier tribe of Waregla. Here several Mistle Thrushes were shot together with Common Cuckoo and Great-spotted Cuckoo. The Great-spotted Cuckoo *Clamator glandarius* was of particular interest to the three remaining members of the expedition and at Kef a considerable number of this species was to be found. Most of the eggs which they accumulated were obtained by their Arab servants so that it was not possible to be certain how they were found in relation to the various nests. Tristram doubted that this species was parasitic on other birds and his views are expressed in Hewitson (1859). It is now known that this cuckoo is parasitic and may lay more than one egg in a host's nest; other cuckoos may lay in the same nest and this has been clearly shown by Martinez et al., (1998), who took blood samples from both adult birds and chicks, so obtaining genetic markers with which to assign parentage. On 27 March Tristram shot a Blue Rock Thrush, one of the few birds that he collected during this part of the expedition.

Journeying on to Souk Harras they spent a few days reconnoitring and added some interesting items to the collections. A Bonelli's Eagle was picked up from a pile of rubbish in the street and the French commandant sent them an enormous Griffon (Vulture) with its wing broken. They watched Barbary Falcons and Red Kites (Plate 29), circling vultures and Lammergeyers, and made their first "oological triumph of the season", in the form of a single egg of the Griffon Vulture. Simpson climbed down to the nest.

Two days after, Tristram set out for Calle on the Mediterranean coast with one Arab servant. Passing the site where together they had watched the Lammergeyers, Tristram entered a more wooded area and watched an Imperial Eagle alight near its nest, which was situated on a horizontal branch in a large oak growing out of the ridge. The nest was made entirely of sticks and about a yard wide. It contained two eggs which were poorly marked with only a few red spots and Tristram added these to his collection. Beyond Hammam Weled Zeid they stopped for lunch during which Tristram watched and listened to Moussiere's Redstart, a bird he described as "an indisputable indigene", since his observations had suggested that the species was not even partially migrant. Up to this point in the expedition, Tristram had not found the nest of this species, though he had collected eight skins. The female he described as having a "modest and inconspicuous plumage" and commented on seeing ten males for every female. In its behaviour he, like Salvin, thought it more of a chat than a redstart. After searching for some time, he found the nest "with a slight skeleton of very small twigs" lined with grass, wool, cows' hair, camels' hair and feathers, mainly those of the Hoopoe; "within this was a very neatly laid lining of fine hair." The nest was very like that of a Redstart, the eggs white, with a faintish tinge of blue-green.

Continuing through a rich and varied landscape dominated by cork oak they entered upon a flower-covered plain in the midst of which was the redoubt of Bou Hadjar. This was a frontier post close to the Tunisian border and manned by seventy troops, two officers and a doctor. Tristram and his servant were the first visitors they had welcomed in the three years since its establishment. The doctor accompanied Tristram into the nearby hills and shot for him two fine male Numidian Magpies (now recognized as a sub-species of the Common Magpie *Picus p. mauritanica*). Neither specimen is included in Tristram's Catalogue, so at some stage they were probably exchanged; this probably happened to many other skins.

After staying the night at Bou Hadjar, Tristram continued on his way, stopping only to watch a Booted Eagle, which at first he took to be a Black Kite. Breakfasting with a Sheik he encountered along the way, Tristram negotiated with him the acquisition of eggs of the Egyptian and Griffon Vultures which he would collect on his return journey. Continuing on his way through country which "must now be very like what Britain was before the Roman invasion" Tristram had what he described as "my first and probably my last rencontre with a lion!" He had spotted a pair of "White Vultures" (probably Egyptian Vultures) circling, and dismounted to search for the nest. Entering some scrubland he came face to face with a young lion and discharged both barrels of his gun at it before he realized it was not fully adult and was probably in the company of its mother who might not appreciate his actions. The young lion fled one way and Tristram in the opposite direction "as fast as the brushwood would permit" to more open and more safe ground. Fortunately there was no sign of the lioness! This was yet one more small adventure for the intrepid ornithologist.

Reaching La Calle, described by Tristram as "a frontier town of French Africa", he spent two days investigating the lakes which lay inland from the town. There were plenty of duck "but very wild, as might be anticipated where French chasseurs were at hand". Pochard, Gadwall, Mallard, Shoveler, Teal and Ferruginous Duck were common but the only herons were the Squacco Heron and the Cattle Egret. With the possibility of being shot by the chasseurs, Tristram "resigned all designs upon the feathered inhabitants of La Calle".

On the way back to Bou Hadjar the Algerian Jay (now considered conspecific with the Eurasian Jay) and Great-spotted Cuckoo were his only captures and he carefully avoided paying a second visit to the lions. Passing again through the oak forest, Green and Great-spotted Woodpeckers abounded and a few Lesser-spotted Woodpeckers were observed together with newly-arrived Rollers. In the evening Red-necked Nightjars flitted about his tent and Scops Owls called throughout the night. At dawn they set out again, obtained a single egg of the Egyptian Vulture and watched a pair of Tawny Eagles (Plate 30); throughout the expedition they were unable to find a nest with eggs, though a live young bird was later brought to the camp and was carried back to the Zoological Gardens in London. In his account of his short stay in Bou Hadjar on his return journey, no mention was made of any vultures' eggs collected for him, so presumably none were. From Bou Hadjar Tristram and his servant rode to Kef Laks where Salvin had set up camp and was hard at work. Beyond this point Tristram did not record details of his activities in print but additions to his collections can be deduced from his Catalogue (1889).

At Kef Laks, on the north side of Djebel Dakma, Salvin examined the nest of an Egyptian Vulture from which the only egg had been broken. The nest was more easily approachable than that of the Griffon Vulture, was placed in a crack in the rocks and was composed entirely of small sticks. A pair of Tawny Eagles was also present in the vicinity but they were still without eggs in "about the fourth week of April". The most interesting acquisition from the area was a clutch of three eggs of the Barbary Falcon from a nest found on 20 April by one of the Arab searchers. One of the adult birds was also collected. Also towards the end of April the European Bee-eaters returned to Kef Laks, and several of the nests were later dug out. The tunnels opened gradually from the round entrance hole until reaching a domed end chamber of about a foot in diameter. The clutch, usually of six eggs, was not completed until early June and Salvin interestingly recorded that several nest holes had a side branch from what was apparently the end chamber, extending about a further foot to a second chamber which contained the eggs.

The Grey-headed Wagtail was seen to be locally common on migration, but more attention was given to breeding birds than to migrants. Tawny Pipits were collected at Kef Laks and afterwards the species was found commonly on the plains of Djendeli where eggs were obtained which varied much in markings and colour, some like wagtails, others profusely marked. There was no success here in finding the eggs of the Blue Rock Thrush though the bird was common. Again at Kef Laks, Salvin shot a pair of Bulbuls, which he referred to as Dusky Bulbuls *Ixos obscurus*. Presumably this was the bird referred to by Tristram as the Garden Bulbul *Ixos barbatus* which is now the Common Bulbul. On reaching Souk Harras they first turned their attentions to the vultures. At this stage they were not aware that the young Lammergeyers had been hatched for some time and, having found some nests inaccessible, continued their search, unsuccessfully. Both the Short-toed Eagle and Black Kite were found to be more numerous around Souk Harras than in any other place the expedition visited. Also here a specimen of the north-African Greenfinch was obtained, shot by Salvin, and nests were later found near Djendeli; it is now considered to be conspecific with the Common Greenfinch but is recognized racially as *Carduelis chloris aurantiiventris*. Of particular interest here was the collecting of a Slender-billed Curlew *Numenius tenuirostris*, shot by Tristram on 3 April, supposedly near Constantine.

On 4 April the expedition pitched their tents at the foot of the rocks of Djebel Dekma and here the Lammergeyer was more accessible. One of the Arab servants was successful and brought down a half-fledged young one, "which, after living for some days came to an untimely end"—and is now in Norwich Museum. It was concluded that this vulture was an early breeder, laying in January or February. Here the birds were feeding principally on the land tortoises, which were very common in the area. Picking them up the vultures carried them high into the air and, often watched by Tristram and Salvin, then dropped them to break the carapace and make the flesh accessible.

The whole southern precipice of Djebel Dekma was the property of a pair of Golden Eagles and they allowed no other raptors to nest there. Here their eggs are usually well marked but the spots more isolated than in Scottish clutches. Salvin remarks that the species built in trees as is the case in Scotland. Only in this area was a pair of Spotted Eagles definitely recognized as such and whilst pairs of Booted Eagles were seen from time to time, no nest was located. They had more luck with the Short-toed Eagle; a nest, with two eggs, was located at Blad el Elma, a village to the south of Djebel Dekma and one of the eggs had slight indications of colouring, unusual in this species. At this camp Mohamed, the expedition's Arab climber, obtained a clutch of three eggs of another pair of Barbary Falcons from a hole high on the rock of Djebel Dekma. Whilst the five-hour climb to the nest was in progress, Salvin shot one of the adult birds. The cave at Djebel Dekma was the haunt of many Jackdaws and also Rock Doves. From the trees on the cliff side Salvin shot a Great-spotted Woodpecker, and on the Medjerdah, just below Djebel Dekma, he shot a Garganey, on 4 April and later, on 9 April, a Tawny Pipit.

Moving on to Khifan M'sakta a nest of the Long-legged Buzzard was found, the eggs being indistinguishable from those of the Common Buzzard. This species was found to be relatively uncommon in areas "where other rapacious birds abound". Simpson shot a single individual near the salt lake of Guerah el Tharf. Red Kites were common in this area and nested usually in trees growing out of the rock though one clutch of eggs was taken from a hole in the face of the cliff. The eggs of Red Kites in Tunisia were almost unmarked. In a wooded ravine Salvin observed Nightingales in the early part of April and he commented that in the salt-lake districts the species does not seem to occur. Cirl Buntings were also found around Khifan M'sakta but, as they were rapidly moving on between nightly campsites, recording took precedence over collecting, though a Great-spotted Woodpecker and a Green Woodpecker were shot on 15 April, supposedly at Constantine. Striking camp here the expedition proceeded to Kef Laks, where they stayed until the end of the month of April. This was one of the best sites for raptors that they had encountered and they spent a profitable few days there, with Tristram collecting a Yellow Wagtail *Motacilla flava* on 22 April.

Ain Beida, another military station some fifty miles south east of Constantine, was their next objective. Here, on an elevated plain which is an extension of the Atlas Mountains, they camped near salt lakes and apparently did little collecting. Eventually they "gradually felt, as it were, our way westward to Djendeli" where they passed the remainder of the month of May. Here Tristram shot a Little Bustard before they moved on to their final camp near a small marsh west of the road between Constantine and Ratna. From this camp they parted with "every disposable thing" and at Philippville took the coastal steamer to Algiers, leaving North Africa on 11 July.

— 7 —

# A Darwinian Conversion

RAVEN. (*Corvus corax*.

In the middle of the nineteenth century the ornithology of North Africa was little known and Tristram was expecting to find species of birds which were previously undescribed. In this he was not entirely disappointed. His first publication on the birds of North Africa appeared in Volume I (1859b) of *Ibis*. Tristram's paper described nine "apparently new species of Birds":

1. Salvin's Lark *Certhilauda salvini*, Tristram.
2. Long-billed Crested Lark *Galerida macrorhyncha*, Tristram.
3. Sand Lark *Galerida arenicola*, Tristram.
4. Reboud's Lark *Calandrella reboudia*, Loche.
5. Desert Fantail *Drymoica straticeps*, Tristram.
6. Desert Warbler *Sylvia deserticola*, Tristram.
7. Bush Chat *Saxicola philothamna*, Tristram.
8. Salt-loving Chat *Saxicola halophila*, Tristram.
9. Solitary Chat *Saxicola homochroa*, Tristram.

The science of systematics can be said to have begun with the publication of *Systema Naturae* by Karl von Linne (Carolus Linnaeus) in 1758 and was characterized by the central position of the species (Fisher, 1954), which was almost always defined on purely morphological grounds and described from a single or very few specimens. This was the case in Tristram's time but over the 150 years since then, with an increasing understanding of the evolution of organisms, the species has become defined more on biological grounds and the accent has been placed on the population rather than the individual. It is therefore unsurprising that several of Tristram's species are not now recognized. What is of prime importance, however, is that Tristram recognized differences between these birds and those that had previously been described. Only one, Tristram's Warbler *Sylvia deserticola*, Tristram 1859, remains as a recognized species though four retain sub-specific status: Crested Lark *Galerida cristata macrorhyncha*, Tristram 1859; *G. c. arenicola* (Tristram, 1859) (Plate 31); the Mourning Wheatear *Oenanthe lugens halophila*, (Tristram, 1859); and the Desert Wheatear *Oenanthe deserti homochroa*, (Tristram, 1858). *Saxicola philothamna* was sunk as a synonym of the previously described *Oenanthe moesta*, the Red-rumped Wheatear, but retains the English name Tristram's Wheatear as an alternative in *Birds of the World* (del Hoyo et al., 1992). *Certhilauda salvini* and *Calandrella reboudia* are now regarded as synonyms of *Calandrella brachydactyla*, the Short-toed Lark and *Drymoica straticeps* as a synonym of *Scotocerca inquieta*, the Scrub Warbler.

Tristram was probably in too much of a hurry to publish his "apparently new species of Birds" and had he properly consulted his friend Alfred Newton on their specific status at the time he would almost certainly not have described so many as new. In his second paper 'On the Ornithology of North Africa', published later that year (Tristram, 1859c), he wrote of *C. reboudia*: "most probably merely a desert form of the congeneric *C. brachydactyla* (Short-toed Lark)" and of *C. salvini*: "I am aware that it may be termed a local name more properly than a species."

Species or not, this paper had much more importance than the report of possible new species in North Africa. Newton (1896) draws attention to what is almost certainly the first statement in print in support of Darwin and Wallace (1858) and, of course, the later *On the Origin of Species* (Darwin, 1859). This paper by Tristram (1859c) published in the very first volume of *Ibis*, on his travels in North Africa, outlines his acceptance of natural selection:

> Writing with a series of about 100 Larks of various species from the Sahara before me, I cannot help feeling convinced of the truth of the views set forward by Messrs. Darwin and Wallace (1858) in their communication to the Linnean Society, to which my friend Mr A. Newton last year directed my attention, 'On the tendency of species to form varieties, and on the perpetuation of Varieties and Species by natural means of selection'.
>
> It is hardly possible, I should think, to illustrate this theory better than by the Larks and Chats of North Africa.
>
> In all these, in the congeners of the Wheatear, of the Rock Chat, of the Crested Lark, we trace gradual modifications of coloration and of anatomical structure, deflecting by several gentle gradations from the ordinary type; but when we take the extremes, presenting most marked differences. Are these extremes, it may be asked, further removed from each other than the Guinea Negro or the Papuan is from the typical Caucasian? And are these species aboriginal and indigenous or are they developed by climatic and other local

causes? I think the latter alternative almost demonstrable in the case of these birds. These differences of structure (I am using the word here in its widest sense, to include colour, form and size) doubtless have a very direct bearing on the case or difficulty with which the animal strives to maintain its existence. In the desert, where neither trees, brushwood, nor even undulation of surface afford the slightest protection from its foes, a modification of colour which shall be assimilated to that of the surrounding country is absolutely necessary. Hence, without exception, the upper plumage of every bird, whether Lark, Chat, Sylvian or Sandgrouse, and also the fur of the small mammals and the skin of the Snakes and Lizards, is of one uniform isabelline or sand colour. It is very possible that some further purpose may be served by the prevailing colours, but this appears of itself a sufficient explanation. There are individual varieties in depth of hue among all creatures. In the struggle for life which we know to be going on among all species, a very slight change for the better, such as an improved means of escaping from its natural enemies (which would be the effect of an alteration from a conspicuous colour to one resembling the hue of the surrounding objects), would give the variety that possessed it a decided advantage over the typical or other forms of the species. Now in all creatures, from Man downwards, we find a tendency to transmit individual varieties or peculiarities to the descendants. A peculiarity of either colour or form soon becomes hereditary when there are no counteracting causes, either from change of climate or admixture of other blood. Suppose this transmitted peculiarity to continue for some generations, especially when manifested advantages arise from its possession, and the variety becomes not only a race, with its variations even more strongly imprinted upon it, but it becomes the typical form of that country. If it be objected that we see many varieties that do not become hereditary, we may reply, that these varieties have experienced changes not advantageous to their means of existence, may from that very cause become extinct.

Still there are many that continue, as the Pied Raven of the Faroe Islands and the Tailless Manx Cat.

To apply the theory to the Sahara. If the Algerian Desert were colonized by a few pairs of Crested Larks—putting aside the ascertained fact of the tendency of an arid, hot climate to bleach all dark colours—we know that the probability is, that one or two pairs would be likely to be of a darker complexion than the others. These, and such of their offspring as most resembled them, would become more liable to capture by their natural enemies, hawks and carnivorous beasts. The lighter-coloured ones would enjoy more or less immunity from such attacks. Let this state of things continue for a few hundred years, and the dark-coloured individuals would become exterminated. The light-coloured remain and inhabit the land. This process, aided by the above-mentioned tendency of the climate to blanch the coloration still more would in a few centuries produce the Galerida abyssinica as the typical form. And it must be noted, that between it and the European G. cristata there is no distinction but that of colour.

But when we turn to Galerida isabellina, G. arenicola and G. macrorhyncha [Plate 32]—now all considered as sub-species of G. cristata—we have differences not only of colour but of structure. These differences are most marked in the form of the bill. Now to take the two former first. G. arenicola has a very long bill, G. isabellina a very short one; the former resorts to the deep, loose sandy tracts, the latter haunts the hard and rocky districts. It is manifest that a bird whose food has to be sought for in a deep sand derives a greater advantage from any elongation, however slight, of its bill. The other, who feeds among stones and rocks, requires strength rather than length. We know that even in the type-species, the size of the bill varies in individuals, in the Lark as well as in the Snipe. Now, in the desert, the shorter-billed varieties would undergo comparative difficulty in finding food where it was not abundant, and consequently would not be in such vigorous condition as their longer-billed relatives. In the breeding season they would therefore have fewer eggs and a weaker progeny. Often, as we know, a weakly bird will abstain from matrimony altogether. The natural result of these causes would be that in the course of time the longer-billed variety would steadily predominate over the shorter, and in a few centuries they would be the sole existing race, their shorter-billed fellows dying out until that race was extinct. The converse will hold good of the stout-billed and weaker-billed varieties in a rocky district.

Here are only two causes enumerated which might serve to create as it were a new species from an old one, yet they are perfectly natural causes, and such as must have occurred, and are possibly still occurring. We know so very little of the causes which in the majority of cases make species rare or common, that there may be hundreds of others at work, some even more powerful than these, which go to perpetuate and eliminate certain forms "according to natural means of selection". But even these superficial causes appear sufficient to explain the marked features of the Desert races which frequently approach so very closely to the typical form, and yet possess such invariably distinctive characteristics, that naturalists seem agreed to elevate them to the rank of species. The differences in size may be yet more simply explained by the facility or difficulty in sustaining existence in varying localities. On similar principles we may account for the existence of such a bird as Galerida macrorhyncha in the warm, genial climate of the Oases, where, winter being unknown, and food always abundant and close at hand, every stimulus is afforded to a vigorous development, while its prey being generally hidden in the soft open mould of the gardens and barley patches, any tendency to the elongation of the bill is fostered and encouraged, until we find a race two inches longer than Galerida isabellina and with a bill exactly double the length (1 inch instead of 0.5 inches).

A process precisely similar may be supposed to have developed the various species of Desert Chats, until we find in the desert of Souf that all distinctive trace of colour has been scorched out, and instead of the brightly clad Saxicola stapazina, we have no more cheerful a representative of the genus than S. homochroa. Widely as these two extremes appear to be separated, yet a well chosen series of the class will exhibit a range of transitions so imperceptible, that it will be found very difficult without careful comparison to draw a line between one species and the next.

I cannot but hope that ere long ornithologists will systematically recognize, what is already admitted in a great degree by conchologists, the clear distinction between species and race. I do not see any great difficulty in taking as a true definition of a species all the individuals who may reasonably be presumed to have a common origin, though among them there may exist races differing from one another even in a considerable degree.

I do not mean for a moment to imply that such birds as *Rhamphocoris clot-bey* (Thick-billed Lark) have been developed out of any known European form, or that we are to presume so far to limit Creative Power as to endeavour to explain the growth of Desert species universally by the development of individual peculiarities. Wherever may have been the centre whence they sprung, undoubtedly there are many creatures to be found there which could not have been developed by any conceivable process from other known races. But whilst it is contrary alike to sound philosophy and to Christian faith to doubt the creation of many species by the simple exercise of Almighty volition, still knowing that God ordinarily works by natural means, it might be the presumption of an unnecessary miracle to assume a distinct and separate origin for many of those which we term species. We may speculate on the question for a lifetime; this conclusion alone so far is certain, - that every peculiarity or difference in the living inhabitants of each country is admirably adapted by the wisdom of their beneficent Creator for the support and preservation of the species. (This quotation is reproduced with the permission of the British Ornithologists' Union <www.bou.org.uk>.)

Whilst the last paragraph somewhat qualifies what he wrote before, he certainly came down on the side of Darwin and Wallace in his analysis of the morphology of desert larks and chats. However, he cannot have anticipated the adverse reaction that resulted from the comments that he made in this paper. Few had the courage "publicly to recognize and receive the new and at the time unpopular philosophy" (Newton, 1896), and this undoubtedly was to affect Tristram's future position in the church.

Newton clearly understood the principle of natural selection in relation to the drab and sand-coloured larks and chats (and the mammals and reptiles) but considered that the black desert forms, particularly the Wheatears, remained unexplained. Tristram later suggested to him (Newton, 1888)—but how much later is not noted—that they escape observation by resembling dark spots and shadows on the rough desert surface. Newton was impressed by the plausibility of this explanation and it was clear that Tristram was still thinking here along Darwinian lines (Plates 33 and 34).

Having once accepted Darwin's evolutionary views, Tristram began to interpret others of his observations in the same way, even speculating on the origin of the Touregs' white dromedaries. "The Saharans maintain the 'mahari' to be a distinct species but it is not necessary to be an acceptor of Mr Darwin's theory to believe that this noble creature is simply the development of the camel by a long course of artificial selection in a very dry, hot climate." Here Tristram was presumably comparing two forms of the dromedary and pointing out that artificial human selection, as opposed to natural selection, may well also result in "the transmutation of species" which was certainly seen to be contrary to religious orthodoxy at the time.

The adverse reaction to Tristram's paper (1859c) to some extent affected the content of his next public pronouncements in relation to his initial Darwinian views. These came in his Presidential Address to the Tyneside Naturalists' Field Club on 29 March 1860 i.e., immediately after the publication of *The Origin*, where he expressed some hesitation in accepting what he termed "the Doctrine of Mr Darwin". A particular point he made concerned the limited fossil evidence, on the grounds that this provided information on the first and last fossils of certain forms but did not "exhibit . . . genealogies of development". He goes on to argue that there is no "need to accept the doctrine of immutability of species against evidence that species may vary under artificial selection, and therefore might and probably would do so under the process of natural selection." (Fig. 3)

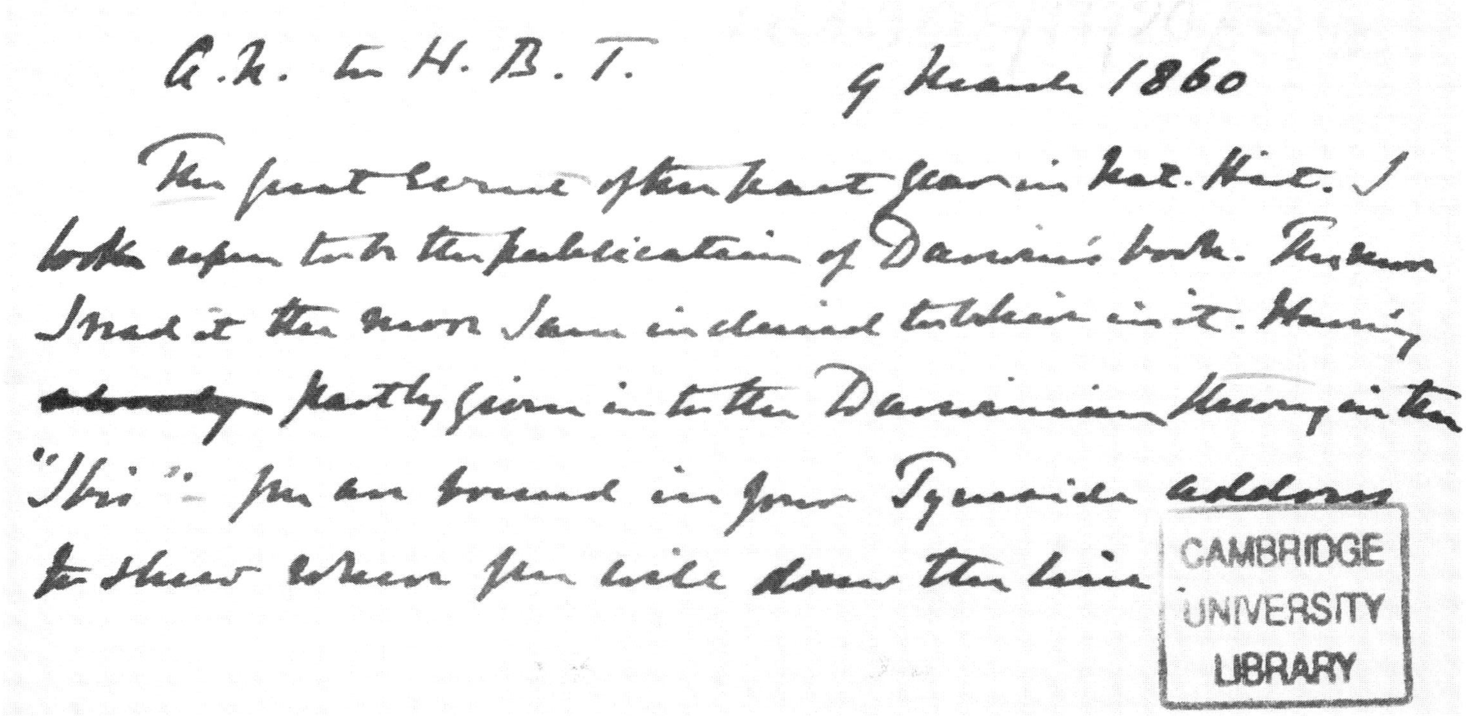

A.N. to H.B.T. 9 March 1860

The great event of the past year in Nat. Hist. I look upon to be the publication of Darwin's book. The more I read it the more I am inclined to believe in it. Having partly given in to the Darwinian theory in the "Ibis" – you are bound in your Tyneside address to shew where you will draw the line



Figure 3: Newton's note to Tristram before the latter's Tyneside Presidential Address to the British Association.

Both Tristram and his friend Newton were very much attracted by Darwin's ideas and the former further expressed his views in his Tyneside address:

> His [Darwin's] work is the result of long-continued thought and labour, directed by a naturalist of extended attainments and remarkable ability, to consolidate a theory which has some facts on which to rest. Some of his postulates we must allow. All organic beings are liable to vary in some degree, and tend to transmit such variations to their offspring. All, at the same time, tend to increase at a very rapid rate, and their increase is kept in check by the incessant competition of other individuals of the same species, or that of individuals of another species, or by injurious physical conditions. Whatever variation occurs amongst the individuals of any species of animals or plants which is advantageous for their existence, will give these individuals an advantage over their fellows, and will probably be inherited by their offspring. It is thus that Mr Darwin assumes all species to have originated.
>
> I feel tempted to say a few words on this, especially as before the publication of Mr Darwin's work, I had expressed this opinion in a paper published in the "Ibis" [1859c], as to many species of birds which I would rather term local *varieties*. But Mr Darwin maintains that the distinction between species and varieties is an arbitrary one and challenges his opponents to say wherein the difference consists. ... Generally speaking I consider that there would be no difficulty in the differentiation of [a] species as an hereditary organism, distinguishable from all others, which either will not hybridise or of which the hybrids are sterile.

Further Tristram makes it clear that he believes that "the application of scripture, as an external authority" should not "be brought to bear so as to check philosophical investigation" and hints at the unreasonableness of questioning what methods any possible creator might use. On the whole the address indicates a degree of support for Darwin and Wallace which is somewhat limited by orthodox religious views.

Tristram sent a copy of his Presidential Address to Newton in the form of two newspaper cuttings and received a reply on 9 April 1860 (Letter CNP 9839/1T/208): "It seems to me that you have been fully equal to the solemnity of the occasion and I congratulate you heartily on the manner in which you propound your datum (starting point). The most pious believer in Darwin must admit that your criticism, even if unfavourable, is fair." At this time Newton was, by his own admission, "in a state of transition". Writing later to Tristram after the Oxford meeting he goes on "but Darwin*oid* I might have remained for a whole geological aeon had it not been for the comments of Bishop Wilberforce during the Oxford debate." (Letter CNP 9839/1T/209)

Based on his paper in *Ibis* (1859c), his Presidential Address to the Tyneside NHS (Tristram, 1860c) and Newton's (1888) recollections, Tristram's position in the Darwinian argument was unequivocal so that it came with some surprise to his friends, and particularly to Newton, that within the year he supposedly recanted. This became public at the July 1860 meeting of the British Association in Oxford (Wollaston, 1921), later known as the Oxford Debate.

The meeting was held in the new University Museum "into which more than a thousand people crowded and hundreds more were turned away"; others watched the proceedings through the windows (Bryson, 2003). The main purpose of the meeting was a paper by Professor J. W. Draper on 'The Intellectual Development of Europe in the light of Darwin's new theory'. Draper's paper seems to have been largely ignored and following this, and after what the chairman Professor Henslow regarded as mainly irrelevant comments, he called upon the Bishop of Oxford to speak.

There is no proper record of exactly what the Bishop said and no official record of the meeting, but it is not true to say that there are no accounts of it. Representatives of the *Manchester Guardian*, *Evening Star*, Jackson's *Oxford Journal* and *The Athenaeum* were all present at the meeting, reports appearing in all four and in three separate issues of the last. In addition there are several accounts (Wollaston, 1921) of correspondence between people present at the meeting and three of these, Professor Alfred Newton (25.7.1860), the Rev. J. D. Hooker (?.7.1860) and T. H. Huxley (9.9.1860), who was one of the main protagonists in what followed, were all highly respected academics. They were very likely to have given an accurate account of proceedings, and on their correspondence the following account is based.

The scene for the 'Debate', which it was not, but which it has gone into history so named, was set two days before on the previous Thursday. In a meeting that day, Owen and Huxley had vigorously debated the possible origins of 'man' from apes and this was to carry over into the British Association meeting. Richard Owen, an anti-Darwinian at this time, who should have taken the chair at the BA meeting on the Saturday, asked the Rev. J. T. Henslow, another academic who was pro-Darwin, to replace him. It is alleged that he did this with the intention of a subsequent trouncing of the Darwinians being magnified by the presence of a pro-Darwinian chairman. Henslow, completely ignorant of the alleged real reasons for the request, agreed to chair the meeting. A small faction, the composition of which is not known, invited William Wilberforce, Bishop of Oxford (known to Newton and his associates as 'Soapy Sam' as a result of Disraeli once describing him as "unctuous, oleaginous (and) saponaceous") to speak at the meeting. Wilberforce, an anti-Darwinian was known as one of the greatest public speakers of his day and was clearly the main attraction for the large audience assembled in the Museum. Huxley, very pro-Darwinian, was also invited to speak. Clearly this was designed to overshadow the main business of the meeting and to be a significant defeat for the Darwinians in the absence of Darwin who was ill at the time.

Accepting the Chairman's invitation, Bishop Samuel Wilberforce gave what Newton (letter to E. Newton, 25 July 1860 in Wollaston, 1921) described as "a wonderfully good speech if the facts had been correct". Newton continued that the Bishop "had made so ill an use of his wonderful speaking powers to try and burke, by a display of authority, a free discussion on what was, or what was not, a matter of truth", and that Huxley reminded him "that on matters of physical science 'authority' had always been bowled out by investigation, as witness astronomy and geology. He then caught hold of the Bp.'s assertions and showed how contrary they were to facts, and how he knew nothing about what he had been discoursing on."

Newton obviously thought that Huxley had won this argument easily but Wilberforce was not without support, even from the scientific community. Richard Owen was Superintendent of the Natural History section of the British Museum and not only did he support Wilberforce but also coached him in the scientific aspects of his verbal arguments and writings. Professor Adam Sedgwick, a leading geologist, also supported Wilberforce whilst several eminent liberal churchmen backed Huxley. This then was not really a debate of science versus religion but of traditional Anglicanism versus liberal Anglicanism, and in the meeting Tristram would far rather have found himself supporting the former. His own field observations conflicted with his normally conservative stance and he was in a quandary.

Perhaps the meeting is best remembered firstly for Wilberforce supposedly asking Huxley whether he claimed his descent from a monkey through his grandfather or through his grandmother (this arising from Huxley having said in the argument of the previous Thursday, "that it was not signifying whether he was descended from a Gorilla or not" (Newton's words) and, secondly, for Huxley's concluding comment that he was not ashamed to have a monkey as an ancestor but would be ashamed to be connected to a man who used his great gifts to obscure the truth. This last statement is reported to have had a tremendous effect on the audience and Lady Brewster is said to have fainted! It is likely that it was also too much for Tristram whose firm traditionalism temporarily gained ascendancy over his own logical thinking as he found himself apparently supporting the Bishop. Newton in the same letter to his brother commented that "Tristram ... waxed exceeding wrath as the discussion went on and declared himself more and more anti-Darwinian."

Darwin was absent from this meeting through ill health and he too suffered the same conflicting emotions as Tristram. Thus it was that Darwin missed Admiral Robert Fitzroy's contribution to the proceedings. Whilst brandishing a very large bible he is reported to have said: "I believe that this is the Truth and had I known then what I know now I would not have taken him (referring to Darwin) aboard the Beagle." (Green, 1996)

In the meantime, in the audience Hooker had become more and more irritated and, in his own words, in a letter (DCP 2852) to Darwin wrote: "I handed in my name to the President as ready to throw down the gauntlet ... Then I smashed him [Wilberforce] and raised applause. I proceeded to demonstrate 1) He had never read your book [*The Origin*], 2) He was absolutely ignorant of Botanical science ... Sam was shut up and had not one word to say in reply ... the meeting dissolved."

Perhaps it was as well that Darwin was not there though it is reported (Ruse, 2001) "that everybody enjoyed himself immensely and all went cheerfully off to dinner afterwards."

Tristram, for no obvious reason other than that he was offended by the treatment of an ecclesiastical colleague, chose to support the Bishop and declared himself not "more and more anti-Darwinian" in the terms that Wollaston took his comments to mean but more probably along the lines of not wishing to be aligned with people addressing a clerical colleague in such a manner. It is not difficult to see how he justified this position which, according to Wollaston (1921), was a source of irritation to his close friend Newton over a period of many years. There is no evidence of any serious disagreement between Tristram and Newton particularly since the two had discussed the content of Tristram's paper at length before it was even written. This discussion took place at Castle Eden in 1858, when Newton called in on his way back from a visit to Iceland. He had made this visit with John Wolley to enquire into the recent extinction of the Great Auk but at this time Tristram had not read the Darwin/Wallace 1858 paper. Newton later showed this to Tristram and must have considered that he had had some small input to Tristram's 1859 paper, so that the latter's apparent recantation of his Darwinian view he had previously expressed would have been doubly disappointing, had it been the case. As a conservative member of the clergy, this was a time for Tristram to keep a low profile on the matter and in so doing he undoubtedly provoked comment from Newton which was little more than banter.

Tristram and Newton were close friends and often made comments to each other to provoke a reaction. Such is the comment from Tristram to Newton which Wollaston (1921) quotes as evidence of Tristram's total rejection of Darwinism: "The more I look into this renovation of Lamarck the more I see it as one blind end" (Letter dated 31 July 1860—in Wollaston, 1921).

Tristram would know very well that natural selection was no renovation of Lamarck and that this taunt would produce a sharp riposte from Newton. In fact Tristram, only four months earlier, stated in his Tyneside Presidential address that Lamarck's "crude theory had been committed to the limbo of forgotten things". Tristram was a traditional Anglican and it is likely that when he saw the lines drawn up at Oxford and the treatment that Wilberforce, a Prince of the Church, received at the hands of Huxley and Hooker, he decided to re-establish his conservatism at least in public. However, privately at first, and later (1868) more publicly in correspondence with Darwin, it is evident that his desert larks were never far from his mind.

It is interesting to speculate whether the Church viewed Tristram's supposed recantation as a true reconversion or whether he was regarded by some clerics as doctrinally suspect. Perhaps the latter was not the case as in 1860 he moved to be vicar of Greatham in Durham and Master of Greatham Hospital which might be considered a promotion, or possibly a small encouragement to keep the old beliefs. However, it would be useful to know whether Tristram's discussion with Newton concerning the chats took place before or after the Oxford meeting of the BA, because if it was after, it would be additional evidence for his having hung on to his Darwinian inclinations privately whilst appearing to reject them in public. Newton, in the meanwhile, developed his acceptance of Darwinism whilst being a regular church-goer, apparently finding no conflict between his science and those parts of the old beliefs that he was prepared to accept (Benson, 1911). For Newton the Bishop's speech during the debate led to his ultimate conversion from Darwin*oid* to Darwinian as he wrote in a letter to Tristram on 30 July 1860—"they caused me by a process of 'natural selection' to become something better. I am developed into pure and unmitigated Darwinism." (Letter CNP 9839/1T/209) This commitment by Newton must have influenced Tristram so that by the time that Darwin eventually contacted him, his initial support for Darwinism was once more firmly established.

It is indeed strange that Tristram and Darwin were unacquainted until the latter wrote in 1868:

Down,
Bromley,
Kent
June 4 1868

Dear Sir,

Although I have not the pleasure of your personal acquaintance, I hope and think that you will excuse the liberty I take in writing to you. I have lately read some papers formerly published by you in the Ibis in which you specify various birds coloured so as to resemble the desert. Now I shd. be greatly obliged if you would inform me whether with these birds the two sexes resemble each other, & whether as far as known, the young resemble the adults. As I did not think that this subject wd specially concern me, I unfortunately returned the volumes without making an extra ch(eck). Therefore I shd be much obliged if you would give me a reference. I read, about a year ago, with lively interest, your work on the Sahara, & if I am not mistaken there was in it a similar description on the colouring of desert birds. If it wd not cause you too much trouble, I should very much like to be permitted to quote from you some such sentence as follows "Mr Tristram informs me that . . . birds, inhabitants of the Sahara, are coloured in a protective manner so as to resemble the surrounding desert. In all these species (or in ⅓ or ½ & CPI) the two sexes in the adult state, & the young resemble each other; yet about ½, ⅓, or ¼ (2) of these birds belong to groups in which the sexes usually differ to a certain extent in colour".

I hope that you will excuse my troubling you & if in your power grant me this favour.

Pray believe me dear Sir,
Yours faithfully,
Charles Darwin.

This letter is written in the hand of Emma, Darwin's wife, but signed by Charles Darwin (information from the Cambridge Darwin Correspondence Project, DCP; see Fig. 4).

Down.
Bromley.
Kent. S.E.

June 4 1868

Dear Sir

Although I have not the pleasure of your personal acquaintance, I hope & think that you will excuse the liberty which I take in writing to you. I have lately read some papers formerly published by you in the Ibis, in which you specify various birds coloured so as to resemble the desert. Now I shd

Figure 4: Darwin's first letter to Tristram penned by Darwin's wife Emma but signed by him (continued overleaf).

inform me whether with these birds the two sexes closely resemble each other, & whether, as far as known, the young resemble the adults. As I did not think that this subject wd specially concern me, I unfortunately returned the vols without making an extract. Therefore I shd be much obliged if you wd give me a reference. I read, about a year ago, with lively interest your work on the Sahara, & if I am not mistaken there was in it a similar discussion on the colouring of desert-birds. If it wd not cause you too much trouble, I shd very much like to be permitted to quote from you

Some such sentence as follows
"Mr Tristram informs me that about .... birds, inhabitants of the Sahara, are coloured in a protective manner so as to resemble the surrounding desert. In all these species (or $\frac{2}{3}$ or $\frac{1}{2}$ &c(?)) the 2 sexes in the adult state, & the young resemble each other; yet about $\frac{1}{2}, \frac{1}{3},$ or $\frac{1}{4}$ (?) of these birds belong to groups in which the sexes usually differ to a certain extent in colour".

I hope that you will excuse me troubling you & if in your power grant me this favour.

Pray believe me dear Sir
  yours faithfully
   Charles Darwin

This is a newly discovered letter from Darwin; the DCP lists it as "not been found". Tristram's reply is very significant. After an initial reply (DCP 6262) dated 6 June 1868, in which he explains that he will be away for three weeks and in order to give a satisfactory reply he needs to consult his books and specimens, he wrote the following (DCP 6234): "I must apologise for my delay in replying to your queries, caused by my absence from home, as your question required the examination of my collections and notes. I have now great pleasure in replying to your kind letter. My views generally were expressed in the Ibis 1859, Vol. 1—pp429.seqq written before your work on the 'Origin of Species.'" Tristram could not have expressed himself more clearly than by drawing attention to what he wrote on page 429 of the 1859 volume of *Ibis*: "It is hardly possible, I should think, to illustrate this theory (ie Natural Selection) better than by the Larks and Chats of North Africa." Seemingly Tristram stood by his initial, pro-Darwinian, statement nine years after he made it and nine years after his supposed recantation.

In DCP letter 6234 and subsequent correspondence dated 5 Sept. 1868 (DCP letter 6351) Tristram provided Darwin with the information that he had requested and this was subsequently published in Darwin's *The Descent of Man* (Darwin, C., 1871, 746) where the date quoted for this is the original 1859 reference:

> Mr Tristram has quoted in respect of the inhabitants of the Sahara, that all are protected by their "isabelline or sand-colour". Calling on my recollection of South America, as well as most of the ground birds of Great Britain, it appeared to me that both sexes in such cases are generally coloured nearly alike. Accordingly I applied to Mr Tristram (letter of 4 June 1868) with respect to the birds of the Sahara, and he has kindly given me the following information. There are twenty six species belonging to fifteen genera, which manifestly have their plumage coloured in a protective manner; and this plumage is all the more striking, as with most birds it differs from that of their congeners. Both sexes of thirteen out of the twenty-six species are coloured in the same manner, but these belong to genera in which this rule commonly prevails, so that they tell us nothing about the

protective colours being the same in both sexes of desert birds. Of the thirteen species, three belong to genera in which the sexes usually differ from each other, yet here they have the sexes alike. In the remaining ten species, the male differs from the female but the difference is confined chiefly to the undersurface of the plumage, which is concealed when the bird crouches on the ground; the head and back being of the same sand-coloured hue in the two sexes. So that in these ten species the upper surfaces of both sexes have been acted upon and rendered alike, through natural selection for the sake of protection; whilst the lower surface of the males alone have been diversified, through natural selection, for the sake of ornament. Here, as both sexes are equally well protected, we clearly see that the females have not been prevented by natural selection from inheriting the colours of their male parents.

There is also an earlier reference in *The Descent of Man* (Darwin, C., 1871, 695) to the same series of letters between Tristram and Darwin:

This way of viewing the relation, as far as it holds good, between the bright colours of female birds and their manner of nesting, receives some support from certain cases occurring in the Sahara Desert. Here, as in most other deserts, various birds and many other animals have had their colours adapted in a wonderful manner to the tints of the surrounding surface. Nevertheless, there are, as I am informed by the Rev. Mr Tristram, some curious exceptions to the rule; thus the male of *Monticiola cyanea* (Blue Rock Thrush) is conspicuous from his bright blue colour, and the female almost equally conspicuous from her mottled brown and white plumage; both sexes of two species of *Dromolaea* (Wheatears) are of a lustrous black; so that these three species are far from receiving protection from their colours, yet they are able to survive, for they have acquired the habit of taking refuge from danger in holes or crevices in the rock.

Reading these letters and Darwin's subsequent commentary on them, it is clear that Tristram appreciated that his observations supported Darwinian natural selection and that Darwin considered Tristram to have accepted his theory. Thus Tristram's support of the Bishop of Oxford, nine years previously, was in no way a rejection of natural selection as Wollaston (1921) considered it to be, but a protest about the manner in which the bishop had been treated. Newton's bantering criticism of Tristram's approach to Darwinism was more probably aimed at his open religious conservatism as a churchman and in not showing more public support for Darwinism. Over the years this was to change.

*Galerida cristata macrorhynca.*

Plate 31: Common Crested Lark *Galerida cristata arenicola*, first described by Tristram as a separate species (*G. arenicola*), but is now considered to be conspecific with *G. cristata* and occurs in the north-east Sahara and north-west Lybia. From Whitaker's *Birds of Tunisia*.

Plate 32: Common Crested Lark *Galerida cristata macrorhynca* is another of the subspecies of the Crested Lark identified originally as a separate species by Tristram, and is to be found in south Morocco and north-west Algeria. From Whitaker's *Birds of Tunisia*.

*Galerida cristata arenicola.*

Plate 33: Mourning Wheatear *Oenanthe lugens halophila* (*Saxicola halophila* of Tristram). The contrasting coloration provides camouflage in stony deserts where stones cast shadows. From Whitaker's *Birds of Tunisia*.

Plate 34: Red-rumped Wheatear *Oenanthe moesta* (*Saxicola philothamna* of Tristram), still referred to as Tristram's Wheatear. Shadows of stones in the desert blend in with the contrasting plumage and provide camouflage. From Whitaker's *Birds of Tunisia*.

Plate 35: Egyptian Goose *Alopochen aegyptiacus*. Described by Tristram as an "occasional visitor", it is now somewhat rarer and an accidental visitor.

Plate 36: Pied Kingfisher *Ceryle rudis*. Still a common resident as it was in Tristram's time.

Plate 37: Osprey *Pandion haliaetus*. A scarce passage migrant in Israel and rare summer visitor.

Plate 38: Goldfinch *Carduelis carduelis*. A common resident and winter visitor to the former Palestine.

Plate 39: Palestine Sunbird *Nectarinea osea*. A common resident which Tristram found particularly attractive. *Ibis* 1865, Vol.6, Plate 2. Reproduced with the permission of the British Ornithologists' Union, www.bou.org.uk.

Plate 40: Tristram's Grackle *Onychognathus tristramii*. The species described and named by Sclater in honour of Tristram and the species by which Tristram is best known. From Tristram's *Fauna and Flora of Palestine* 1884.

Plate 41: European Kingfisher *Alcedo atthis*. A common passage migrant and rare winter visitor.

Plate 42: Nubian Nightjar *Caprimulgus nubicus tamaricis*. Now a much rarer bird than in Tristram's time. From Tristram's *Fauna and Flora of Palestine* 1884.

**Plate 43: Stone Curlew** *Burhinus oedicnemus*. Still a fairly common breeder in the former Palestine.

**Plate 44: Lapwing** *Vanellus vanellus*. Common passage migrant and winter visitor now, in contrast with its rarity in Tristram's time in Palestine. A few breeding attempts have occurred.

— 8 —

# The Long Winters of Palestine 1: South to the Gore (1858, 1863–4)

SCOPS OWL. (*Scops giu.*)

Tristram had much enjoyed his winter sojourn to Algeria in 1855–6 and his health had very much improved as a result of it. So it was that, being not much troubled by his chest in the following winter, he decided that he needed to visit the Mediterranean again in the early months of 1858.

On this occasion he chose to visit Palestine as much to see places and the topography of his Bible lands as to study the natural history (Tristram, 1865c); unreferenced quotations in this and the next chapter are taken from this source.

This was to be the first of six expeditions Tristram made to the then Palestine between 1858 and 1897. In his magnum opus on *The Birds of Israel*, Shirihai (1996) records these visits as "providing the basis for ornithological work in Israel" and states that "his major work *The Fauna and Flora of Palestine* (1884), and a series of articles in *Ibis* (1865c&d, 1866, 1867a and 1868) are amongst the most important sources for comparing changes in the avifauna of Israel." In fact Tristram's first papers on the birds of Palestine appeared earlier in *Ibis* (1859a and 1862), the former being only the second paper to appear in what was to become the foremost ornithological journal in the world. In addition Tristram's literary research was to provide much older information on the birds of Palestine as he himself points out in his *Natural History of the Bible* (1867b). Referring to the title of the book, he writes: "the subject is not without some bearing on the ornithology of the country, as illustrating some of the species most familiarly known and therefore most abundant in Palestine 3500 years ago."

Concerning his first visit to Palestine in 1858, Tristram (1859a) points out: "Ornithology was by no means the principal object of the expedition and the districts most abounding in birds, as the Lebanon, the upper waters of the Jordan, and the wooded regions of northern Palestine were not visited." Despite this comment he ran up a list of 119 species recognized at that time, and ear-marked several places to revisit on future expeditions.

In March and April of 1858 Tristram was collecting in Palestine but his winter cruise on the yacht belonging to the younger brother of Mr Gibbs, Governor of the Bank of England, (Tristram, Louisa H-H., 1898), started much earlier and collecting began in Lisbon and off the west coast of Africa, in Namibia (November). It continued in Tangier, Morocco and Gibraltar, and many places throughout the Mediterranean, including Malaga, Algiers, Tunis (Carthage), Sicily, Corsica, Sardinia, Malta and Zakinthos (January), Navarone and Pylos (February), and Crete and Egypt (March). On 15 March, on the Nile, Tristram collected a Lesser Black-backed Gull *Larus fuscus* at

Boulac, and in April a Francolin *Francolinus vulgaris* (presumably the Black Francolin *F. francolinus*) so that the stay in Palestine could have been only for about three weeks and probably the party followed a typically tourist route.

Tristram's second expedition to Palestine in 1863–4 was a much grander affair. Apart from Tristram himself, W. P. C. Medlycott, H. T. Bowman (Photographer), B. T. Lowne (Botanist), Edward Bartlett (Zoological Assistant), H. M. Upcher and C. W. Shepherd all started the journey; they were joined by J. H. Cochrane, and Messrs. Barneby-Lutley, Garnier and Egerton-Warburton for what Tristram described as "the most interesting part", the trans-Jordanic part of the journey. Subsequently Shepherd and Upcher were to be elected to the then very select British Ornithologists' Union. At the time of the journey the BOU numbered only twenty members. The initial plan was to travel to the Dead Sea and east of Jordan, areas least accessible to travellers and of which there was a very incomplete knowledge.

After landing in Beirut, several days were spent organizing the expedition and taking trips out of the town as far as Rabda. In the region of Beirut, whilst crossing the bay, Gull-billed Terns were common and both Adriatic (Mediterranean) Gull, which Tristram had collected in Malta, Audouin's Gull and the "Herring Gull of the North" (surely the Yellow-legged Herring Gull) passed overhead. In the green areas of Beirut wintering Willow Warblers and Chiffchaffs were regularly encountered. The presence of the little Egyptian Fantail reminded them that England and Europe had long been left far behind though robins and hedge accentors were to be found commonly. The bulbul "the nightingale of Palestine" was plentiful, skulking in the orange groves, white and yellow wagtails abounded, and in the cold streams the water ouzel (Dipper) was to be found. "Thus in our first day's expedition we had abundant illustration of what to the field naturalist is the most marked peculiarity of Palestine, the juxtaposition of northern and southern forms of life . . . within the narrowest limits." (Tristram, 1865c)

The departure of the expedition from Beirut was on the 28 November and the plan was to follow the coast for about eighty miles, as far south as Haifa, and then to turn inland. The shoreline was achieved only after leaving the Beirut peninsula and here they encountered shorebirds wintering from the north, particularly Dunlin, Stints and Redshank. Of particular interest to Tristram was "a rare kind of Wheatear—the *Saxicola libanotica* of Hemprich and Ehrenberg, one of the birds peculiar, as far as we know, to the stony regions of Palestine" (Tristram, 1865c). This Tristram obtained but there is no record of it in his Catalogue of 1889. It is now considered to be a sub-species of the Northern Wheatear *Oenanthe o. libanotica*, and in his *Ibis* paper referring to this expedition Tristram (1867a) comments on its close similarity to *Saxicola lugens*. Both birds are now placed in the Genus *Oenanthe* and the latter, which Tristram refers to as the Pied Chat, is now named the Mourning Wheatear (del Hoyo et al., 1992), the name Pied Chat now being attributable to the Variable Wheatear *O. picata*, as a secondary name in some areas.

The forty or so miles to Sidon, along the coast, produced little of ornithological interest, except the collecting of a Blue Rock Thrush. After setting up camp near Sidon, Tristram took his gun and shot several corncrakes "identical with our own" and some bulbuls. Overhead, birds of prey were common and he identified Honey Buzzard, Marsh Harrier and several eagles which circled out of sight; one Short-toed Eagle, however, fell to his gun but was then lost amongst the hedges. On one excursion from the camp heavy rain set in whilst they were wandering along the banks of the River Auwaly, north of Sidon, and as the weather changed they observed many birds heading up river. Here was the only place in Syria that members of the expedition saw the Pygmy Cormorant and these were followed up river by many "oceanic ducks", and Pied and European Kingfishers flying under the tamarisks for shelter.

Eventually, when it was obvious that the rain was going to continue, the expedition set forth again south and encountered many birds at the first river crossing. At Nahr Seniku eagles, ducks, Egyptian Geese (Plate 35), falcons and plovers surrounded them but few were collected as most of the firearms were damp. At the second river crossing, the Nahr-ez-Zaherany where the ford was close to the river mouth, they had much more difficulty in negotiating it. One of the bearers was nearly drowned and Tristram's portmanteau fell into the river and all the contents were soaked. After camping at Sarepta, after a wet night, on 3 December the expedition aimed south for Tyre and the River Leontes, now Nahr-ez-Kasimiyeh, across rocky ridges where "Black Wheatear" or Tithys' Redstart, now Black Redstart, occurred commonly.

Only a few small birds were collected during the next few days and on 5 December a single Black-shouldered Kite evaded Tristram's gun; this was the only example of the species seen on the expedition. In the more vegetated areas what Tristram termed the Greek Partridge, now the Chukar Partridge, ran along the ground to avoid them, and as they descended into the plain bounded by the Jebel Mushakka, Pied Kingfishers (Plate 36) hunted above the streams. It was planned to camp in the village of El Bussah where the expedition arrived in the late afternoon. Immediately, as the camp was being set up, Tristram sallied forth with his gun and an army of small boys offering to act as beaters. Despite the presence of more than fifty assistants he managed to shoot three Little Owls and a Great Grey Shrike and apparently none of his enthusiastic helpers.

A birdless Sunday was followed on 7 December with a visit to the coastal town of Zib which lies at the mouth of a swift mountain stream flowing through the Wadi Kurn. Here they found a pair of Tawny Eagles. Tristram writes: "We dismounted, and got within easy shot by stalking, but disgracefully we lost our game." Soon after, they came upon a pair of Bonelli's Eagles "who shook their wings contemptuously at the assault of our small shot". Later in the day they watched six circling Griffon Vultures which eventually, having discovered that their potential meal was not yet dead, disappeared into the vault of the sky. Riding back to camp they encountered more Greek Partridges. Tristram mentions in his description the similarity to the Chukar Partridge of India and speculates that though they may well be the same species, there are some characteristic differences. Apparently he found them much better eating than our Red-legged Partridge and in fact dined that evening on them in a stew with Woodcock, a species which he had not mentioned encountering on the expedition. On his way back to camp he put up a Golden Eagle.

The following day, 8 December, was a red-letter day on which Tristram obtained "perhaps the most interesting addition, as well as the most unexpected, which we made to the fauna of the country." (Tristram, 1865d) Still in north-western Galilee, running up to the Plain of Acre, the party was in the wooded glen of Wadi Kurn (Nahal Kesiv), sheltering under a carob tree from which they flushed a large owl. Their first reaction was that it was an Eagle Owl but its long, bare tarsi and Osprey-like talons confirmed it to be the Brown Fish Owl. Previously this species had never been recorded west of southern India and in the next two days they saw three more. Now, sadly, this owl is no longer to be found in what was Palestine where the last confirmed sighting was in 1975. Shirihai (1996) attributes the disappearance of the owl mainly to the drainage of waterways and the extensive development of its habitat, though it was probably very rare even in Tristram's time.

The next day the party set out for Caiffa, at the foot of Mount Carmel, intending to visit Acre on the way. The plain they crossed was

> abundantly stocked with game of every kind. In the lower swampy portion, we heard, though we could not see, the Francolin, once the dainty of Italian epicures, by now utterly extinct in Europe. It is still frequent in Cyprus and in all the lowlands of Syria and is well-known to Anglo-Indians by the name of the Black Partridge. Of plover we found and obtained abundance of many sorts, Golden, Green and Kentish, winter visitors from Europe and the red-throated and Asiatic Dotterels from Eastern Russia. The pretty and lively little cisticole, well-known in Sicily and Algeria, a warbler smaller than our wren, frequently rose lark-like from the tufts of rushes and was added to our collection; and the whole plain was stocked with birds of prey of every kind, from eagles and falcons to harriers and sparrow-hawks. As there was no cover it was difficult to approach, but M. [W. P. C. Medlycott] secured a fine specimen of the Common Buzzard (*Buteo vulgaris*, L.) by the judicious use of an ass as stalking-horse.

The delightful campsite selected for the expedition, in view of the consul's house in Caiffa, proved to be a disaster as they awoke next morning in the middle of a stream, with all their equipment sodden. The consul suggested they move to the convent on Mount Carmel which provided excellent accommodation, and Tristram and Medlycott, after a hearty breakfast, immediately went out collecting, returning with a Redshank and a "Manx

Shearwater". This latter is recorded in Tristram (1884a) as such but does not appear in the Catalogue (1889). It is much more likely to have been the Yelkouan Shearwater *Puffinus yelkouan* which, for a time, was considered to be a subspecies of the Manx Shearwater.

On the return to the convent Tristram and Medlycott found what can only be described as a "takeover". Tristram writes:

> We had a long suite of rooms opening into each other. Next our sitting-room was my chamber filled with saddles and camp properties. Next M's, elegantly furnished with his paints, drawings, tools and shell boxes. Through this we passed into B's [Bowman's], very like a chemist's shop, filled with his photographic apparatus. L's [Lowne's] followed, carpeted with layers of botanical papers and discoloured plants which would not dry; and B-----t's last, with a strong odour of bird flesh, and a long array of disembowelled bird specimens.

It is not surprising that Tristram did "not feel disposed to criticize monastic orders" and realized "the uses and value of religious houses to the traveller in the middle ages"—and one would think also to those travelling in the nineteenth century.

In the convent buildings in Caiffa in which the expedition had set up its semi-permanent laboratories were the remains of the nests of the Red-rumped Swallow though in the time which Tristram's group were there the birds had migrated south. Present, however, was what Tristram called the Oriental Chimney Swallow, which he regarded as of specific status, but is now regarded as a subspecies of the Barn Swallow; this was present in great numbers.

Two days collecting around the mouth of the Kishon produced several species of interest. A Red-breasted Merganser and a Great Black-headed Gull (the Red Sea Gull of Tristram) and after great exertions a male Greater Flamingo was shot in flight by Bowman. Herons and Ospreys (Plate 37) stayed well out of range but Tristram obtained an Audouin's Gull. Some time was spent riding out from the convent as far as Esfia and exploring up the valley of the Kishon. Large numbers of water birds were present here, including waterhens, coots and Little Grebes and for the first time on the expedition they met with natural forests. Here were found Black-headed Jay, Wood Pigeon and the Syrian Woodpecker, the only woodpecker to be found in Palestine, a reflection of the paucity of woodland.

After leaving Caiffa, by 16 December the party had reached Nazareth and as they explored the area on the following day, colonies of Lesser Kestrels were found on the steep cliffs behind the town. Tristram thought Nazareth a very poor place and after walking some time in the surrounding hills he expressed his pleasure at the departure of the group on the morning of 18 December. The route to be followed by the expedition lay due south with Jenin as the next main objective. To begin with their road lay southeast to Endor from where Griffon Vultures could be seen circling over Mount Tabor, gradually rising until they disappeared, leaving the eagles well below them. Little else of ornithological interest was observed before they reached Jenin. There, in the evening, walking in the gardens before the evening meal, they encountered a single owl, the Egyptian Eagle Owl *Bubo ascalaphus*, later regarded as conspecific with the nominate form as *Bubo b. ascalaphus* in Shirihai (1996) but again given specific status as the Pharaoh Owl by del Hoyo et al. (1999), though the DNA evidence is inconclusive. The only other observation here was of a Hooded Crow considered by Tristram to be the same bird as in England but now given sub-specific status as *Corvus cornix pallescens*, being a smaller and greyer bird than those of more westerly populations.

From Jenin the route westerly of Jerba was taken towards Nablus, and along the way a group of Black-winged Stilts (Tristram's Stilted Plover) was found in a temporary pool of rainwater. The highlight of the journey, though, was a fine (Eastern) Imperial Eagle "which came and hovered over our path for some minutes but no gun was loaded for him at the moment. He was a sight that the naturalist rarely sees so closely—jet black with pure white shoulders, and white under the tail, he well deserved his imperial title." Jays and several specimens of the Syrian Woodpecker were not so fortunate and were obtained by members of the expedition. They were welcomed outside the gates of Nablus by woodpeckers, jays and jackdaws calling from the olive groves.

Whilst Tristram considered Nablus to be the best town that they had encountered since leaving Beirut, its birds went unrecorded and it was not until they were approaching Shiloh on 22 December that any collecting took place. Here they again encountered the chats of the northern rocky hills and the Black Redstart. Amongst the chats Tristram spotted one bird which was quite distinct and he eventually caught up with it. He identified this as the Arabian chat *Saxicola xanthomelaena* and commented that this was the only occasion on which it was encountered. There is some question about the identification of this bird. It is unrecorded in Tristram (1884) or in the Catalogue of the Collection (Tristram, 1889) either under this name or as *S. melanoleuca* with which it had been synonymised. There is a problem here in that Tristram records this latter bird as not returning to its breeding grounds in Palestine until about 16 March so that it is likely that the bird he identified as *S. xanthomelaena* was not *S. melanoleuca* with which he was familiar. The latter is now regarded as a subspecies of the Black-eared Wheatear *Oenanthe hispanica*. Shirihai (1996) records *O. h. melanoleuca* as a breeding summer visitor, widespread on passage and rare in winter in Egypt and Arabia.

Travelling through an area of fig trees in the valley approaching Bethel (Beitin), Jays, Woodpeckers and Little Owls called incessantly, allowing several trophies to be collected from the bare branches. The plan was to stay in Jerusalem over the period of Christmas, and ornithology was replaced by planning for the journey to the Dead Sea and for visiting the sites of the city. In many areas the group must have seen birds which went unrecorded but at the sacred place of the Haram Wall they were too numerous to ignore. The return to ornithology was stimulated by Tristram finding "the nest of a sparrow of a species so closely allied to our own that it is difficult to distinguish it." Several other species were present in the region of the wall and pairs of the Palm Turtle Dove fed together with the Common Turtle Dove. The latter is migratory whereas the Palm Dove, the most southerly distributed and least abundant of the species which are found in what was Palestine, are sedentary. Goldfinches (Plate 38) and Great Tits were present in the Cypress trees, and Blue Rock Thrushes and White Wagtails on the ground. Kestrels were present in the domes and circling in front of the walls occupied later in the year by Lesser Kestrels and the Scops Owl. However, the characteristic birds of the Haram were the crows which came in to roost at night. "From the solemn Raven down to the impertinent Jackdaw, all were there." "Of all the birds of Jerusalem, the Raven is decidedly the most characteristic and conspicuous." In watching them come in to roost, Tristram noticed a different call note—that of the Brown-necked Raven in amongst the larger Common Raven. Hooded Crows, Rooks and Jackdaws also came into roost each night in their hundreds. Despite a warning from the Consul that any attempt to collect specimens from the region of the wall may be considered sacrilege by the Muslims, the members of the expedition devised a cunning plan. "My companions were anxious to obtain specimens of these Jerusalem birds" wrote Tristram, as though he were not. "My friends determined nevertheless to run the risk (of the muslims' ire) and stationing themselves just before sunset, in convenient hiding places, at a given signal they fired simultaneously, and safely gathering up the spoils had retreated out of reach, and were hurrying to the tents before the alarm could be raised." This was repeated on a second evening but on the third attempt the ravens were too clever and avoided the shot. On the first night "the discharge of ten barrels had obtained fourteen specimens, comprising five species."

In the area outside Jerusalem there were few birds, probably because of the lack of vegetation, but Fieldfare (though no Redwing) were observed, despite the latter reaching Israel in small numbers nowadays (Shirihai, 1996). Of particular interest was the acquisition of a Whooper Swan from the market, shot on Solomon's Pools, near Bethlehem. This was the first record for Israel.

On 30 December the expedition set out east for Jericho and almost immediately Tristram shot what he described as "one of the peculiar birds of Palestine, a pretty black and white chat" which he identified as *Saxicola libanotica*. This was in fact the central Eurasian form of the Northern Wheatear *Oenanthe oenanthe libanotica,* a migratory bird in Palestine. "We obtained several interesting and novel specimens as we walked along, especially a new desert lark—a small bird of rich russet-red plumage and varied note (*Ammomanes fraterculus,* Tristram), not unlike the Isabel Lark of Spain and North Africa, a very graceful little bird, slate-coloured with black tail, of the size of our Robin, and resembling the Stonechat in his habits, which we named the black-tail." Now it is regarded as conspecific with the Desert Lark but recognized as a subspecies *A. deserti fraturculus*, found only in the Dead Sea Depression. "We also found a beautiful little Partridge of the Dead Sea basin, rather smaller than ours with bright orange legs and beak and its flanks striped with black, white and chestnut"—now the Sand Partridge.

Skirting the tremendous gorge of the Wadi Kelt the party was able to look down on ravens, eagles and vultures circling below, and that night they camped on the remains of Jericho. "In zoology Jericho surpassed our most sanguine expectations. It added twenty-five species to our list of birds collected on the tour, and nearly every one of them rare and valuable kinds." Tristram was confined to his tent for two days but the rest brought in more specimens than could possibly be preserved and the soup pot got the benefit.

Tristram's description of the ornithology of the region could not be improved upon:

> The bulbul, or Palestine Nightingale (*Ixos xanthopygius*) positively swarms, almost every tree being inhabited by a pair, and the thickets re-echoing with their music; the comical and grotesque-looking "hopping thrush" as we have named the *Crateropus chalybeus* [now the Arabian Babbler *Argya squamiceps*]—jumps and spreads his long tail in every glade; the gorgeous Indian Blue Kingfisher (*Alcyon smyrnensis*) [now the White-throated Kingfisher *Halcyon smyrnensis*] perches solemnly over the little rivulet; the Egyptian turtle-dove inhabits the taller trees; and various little warblers of Indian or Abyssinian affinity skulk in the thickets. On the plain above are the desert larks and chats, while half-an-hour's walk takes us to the Mount of Temptation (Mons Quarantania), the home of the Griffon, the beautiful little Hey's Partridge, Tristram's Grackle, various rock-swallows and Galilean Swifts, and the wildest of Rock Doves in swarms. But beyond all others, Jericho is the home of the lovely little sunbird (*Cinnyris osea*) [now Palestine Sunbird *Nectarinia osea*, Plate 39] hitherto only known in Europe by Antinori's unique specimen, though mentioned by Lynch, De Saulcy and others as a humming-bird,

a genus exclusively confined to the new world. The male of Hosea's Sunbird is resplendent with all the colours of the humming-bird, and not much larger than most of that tribe, measuring 4¼ inches in length. It has a long, slender and very curved bill, all the neck a brilliant metallic green, the throat metallic blue, and the breast metallic purple, with a tuft of rich red, orange, and yellow feathers at each shoulder (the axillary plume), which he puffs out as he hops in the trees, paying his addresses to his modestly-clad brown-green mate.

Then the grave-looking [Great] grey shrike sits motionless on the topmost boughs, lost in amazement at the proceedings of the howadji in their tents below, or waiting for the passing of some droning beetle; and the merry little long-tailed wren (*Drymoeca gracilis*)—[now Graceful Warbler *Prinia gracilis*]– speads its fan-like tail as it runs up the twigs of the tamarisk. These are only a few of the ornithological riches of Jericho.

By 4 January Tristram was fully fit again and was joined by Bowman, Upcher and Shepherd in a day's exploration of the northern part of the Jericho plain and the caves of Mount Quarantania. First skirting the base of Jebel Keruntil they began to climb and immediately encountered what was soon to become Tristram's Grackle (Plate 40), a bird

well-known to all the visitors of the Convent of Marsaba as the orange-winged blackbird. It is a bird exclusively confined to the rocky gorges round the Dead Sea, and the gorge of the Kedron at Marsaba. It may, perhaps, be found at Petra. Geographically considered, the occurrence of this bird here is very interesting, for it belongs to an exclusively African group, without any representatives in Europe or Asia; and certainly no member of the genus occurs further north than Abyssinia, save this isolated and restricted species. It is considerably larger than our Blackbird, with lustrous black plumage and rich chestnut coloured wings. Its note is of wondrous compass, rich and sonorous—I think the most powerful and melodious whistle I ever heard—as it re-echoes from cliff to cliff. Wild and wary, it lives in small flocks of five or six, and it requires no little perseverance to approach it within shot (Tristram, 1865c).

This, however, Tristram achieved and here he shot a single female and later, in April, a male in the same place, to add to the male and female in his collection shot at Marsaba in March of 1858.

A quarter of an hour later, after scrambling up the slope above them, they came to the base of the cliff which was scarred by scores of caves, many of them man-made and some of them interconnected. Bowman and Shepherd led the way and Tristram, initially wary of the 700 feet drop, with vultures circling below, eventually followed. Hundreds of Rock Doves flew out of their sanctuary as the explorers progressed and they found the roofs of the caves adorned with the used and abandoned nests of the Little (Galilean) Swift, Red-rumped Swallow and Rock (Crag) Martin.

On 6 January, on an excursion to the Jordan, Upcher shot a Spotted Sandgrouse from a passing flock, which was surprisingly far east for the species. Against the advice of their guides they continued collecting along the river and came to no more harm than a soaking for Tristram, who fell off his horse in swampy ground whilst trying to retrieve a duck he had shot! Collecting post in Jerusalem and visiting various archaeological ruins took up most of the following days, though a grackle, sunbirds, bulbuls and desert partridge increased their ornithological stores. For the first time, on 11 January Tristram encountered what he described as Menetrie's Wheatear—now known as the Isabelline Wheatear—"which has a disagreeable habit of sitting on top of a bush, out of gunshot, and then, on the approach of danger dropping down into a burrow of which the plain is full." In an attempt to dig it out he blocked three tunnels but it escaped through a fifth—a yard away. On the way back to camp Tristram was surprised to see, in winter, Galilean (Little) Swifts circling above him and on arriving in camp to discover that in his absence a Collared Dove (now *Streptopelia decaocto*) had been collected, "an Indian and Asiatic species, which we should certainly not have expected to meet with here, certainly not in winter." In the latter half of the twentieth century this species spread north into Europe and is now a common bird throughout the British Isles, first breeding here in 1955.

Proceeding northwards along the west bank of the Jordan, a Desert Wheatear was collected, described by Tristram as "a native of Nubia which had wandered north". Now it is a common resident and passage migrant, and winters in the southerly desert regions. Turning south again, there appeared to be no logic in the wanderings of the expedition. Their advisers seemed intent on keeping them west of the Jordan and they headed for camp again on the banks of the Dead Sea near El Feshkhah. Here a second Collared Dove was shot and on an area where the plain had been flooded the footprints of Black Stork, Redshank and Sandpipers mingled with those of small mammals. Other species of birds were common here, including eagles, ravens, warblers, and both Chukar and Sand Partridges were shot, the latter providing lunch. As they arrived on the shore of the lake Tristram shot a Brown-necked Raven, Upcher a European Kingfisher (Plate 41) and an Audouin's Gull. Along the shore Dunlin and Redshank were feeding together with several wagtails and another Desert Wheatear was obtained. On the lake itself a group of Pochard swam offshore, some distance out. Tristram comments in his account of the day that most organisms do not survive well in the highly saline environment but birds seem to be something of an exception. Where the Jordan poured into the Dead Sea, in a torrent, he recorded that there was little vegetation. Here he shot a Golden Eagle which fell on the wrong side of the river "to waste his carcass on the jackals and vultures in the Land of Moab". During the trek south that day the party passed a point on the Jordan that Tristram had visited six years previously.

Some miles down the shore of the Dead Sea the party struck inland over a marshy area where they found Norfolk Plover (European Stone Curlew, Plate 42), a flock of Black Stork and a solitary Crane, all, fortunately for the birds, out of gunshot. On reaching camp at the point on the shore where the fountain of Ain Feshkhah sends its warm waters, steaming, into those of the Dead Sea, Tristram found that the day had been a prosperous one in natural history, "Bartlett had found a specimen of a very small nightjar, which turned out to be a new species *Caprimulgus tamaricis* (Plate 43) . . . and . . . another new wheatear had been added to our list." The nightjar is now considered to be a sub-species of the Nubian Nightjar but retains tamaricis as the sub-specific name, whilst there is no indication of the Wheatear, either in Tristram's books or his Collection Catalogue. The following day Shepherd collected both the Pied and White-breasted Kingfishers and three more Tristram's Grackles were shot.

Early next day (15 January) the expedition set off for the Convent of Marsaba and almost immediately obtained a specimen of the White-headed Black Chat *Saxicola leucocephala,* first found by Tristram in the Algerian Sahara; there were three birds in the group and no others were seen during the expedition. Now considered to be synonymous with the White-crowned Black Wheatear *Oenanthe leucopyga,* the species is nowadays a common resident of south and east-central Israel.

On 18 January they had expected to return to the Dead Sea but no guards arrived so they collected birds around the convent where they obtained a Blue Rock Thrush, a Crag Martin, a Black Redstart and several desert larks and chats. The following day they headed back east and struck south when they arrived at the shoreline, making camp at Ain Terabeh. The oasis was full of life and the bush dense, which made the birds harder to see, difficult to shoot and nearly impossible to find afterwards. On the water were Pochard, Teal and Great-crested Grebe and in the bush "Hopping Thrush", Bulbuls and what appeared to be a large sunbird, never seen again. However, the most important find turned out to be a species new to science, a very small and richly marked sparrow, now known as the Dead Sea Sparrow *Passer moabiticus,* "the size of a domestic sparrow, with chestnut wings and a rich yellow patch on each side of the neck. The female of uniform russet plumage also exhibits, but less distinctly, the yellow patches. It is indeed strange and interesting to discover, in this little restricted locality, a species which seems strictly confined to its narrow limits, and not a straggler from Africa or India." After this discovery they immediately found the fresh tracks of a leopard which evidently had its lair in the cane brake they had been working over, but nothing daunted, attempted unsuccessfully to obtain a bird from a party of passing Ravens. This was a species with a shorter tail than the common Raven, broader wings and a shrill call, "almost musical". Later they discovered that this bird was the Fan-tailed Raven *Corvus rhipidurus*.

Just before reaching Ain Jidy, in a bay at the base of Jebel Shukif, sulphurous fumes were bubbling out of the ground and the water running through the gravel into the lake was some 33 degrees F. warmer than the lake itself. The ground around was bare of vegetation and there were no birds in the area around, whereas elsewhere they were relatively abundant with ducks and grebes apparently feeding successfully out in the lake. In setting fire to a cane brake in order to make a path, the conflagration got out of control and created a lot of smoke—"a bonfire on Titanic scale"—which brought in a party of Griffon Vultures; none had been seen for some time before the fire and Tristram speculated that they may have been attracted by the possibility of food.

The collectors brought in two interesting items. Firstly, Shepherd had shot a Striolated Bunting, a new bird for the Palestine expedition, which was familiar to Tristram from the Sahara. Secondly, Bowman had shot an emaciated Lapwing (Plate 44), which Tristram regarded as a rare straggler though now it is a regular passage migrant and there have been two records of breeding in the north of Israel (Shirihai, 1996). On the following day, 23 January, Tristram and Upcher decided to explore the Wadi Areyeh where they shot a long-tailed warbler which Tristram thought to be a new bird for the expedition though he commented that it was very similar to the Scrub Warbler *Scotaria inquieta* of the Sahara. At that time this bird was considered to be in the Genus *Drymoeca* and Tristram named it *D. engedensis* initially (Tristram, 1865c) but later *D. eremita* (Tristram, 1867a). Here Tristram comments: "So wild and wary was it, that the first specimen observed, which we at once recognized as a species new to us, cost Mr Upcher and myself two hours pursuit in the Wady Areyeh, and eleven shots, before we secured it." Subsequently it was correctly confirmed as a Scrub Warbler, which is now a common resident in Israel, though it did not extend its present northerly distribution until the 1950s (Shirihai, 1996).

Striking the camp at Engedi the expedition moved on to Masada (Sebbeh) where they found the water supply had dried up. In climbing up to the fortress they encountered flocks of Rock (Crag) Martins and on reaching it at a calculated height of 2,200ft above the Dead Sea they had excellent views. Circling above them was an Imperial Eagle and they watched a Lanner Falcon pursue a flock of Rock Pigeons for some time. Back at the bottom, they found that Shepherd had obtained a Wedge-tailed Raven and the remains of a Pochard taken by a Lanner Falcon. Later Tristram collected what he described as a Nubian Wheatear now known as the Hooded Wheatear *Oenanthe monacha*, which is nowadays a scarce local resident in the Negev. Because of the shortage of water they moved on next day, and with some excitement, Tristram wrote: "At length this morning we leave Palestine proper on our anticipated visit to the east side and the desolate Lisan—or Peninsula—we are to enter the Land of Moab." Little did he know what disappointment awaited them. Heading to the very southernmost point of the Dead Sea, during the progress, Tristram shot another White-crowned Black Wheatear. The lake became much shallower at its southern extremity and in a marshy area Upcher collected a Coot and a Water Rail. Though there seemed little for the birds to feed on, in what appeared effectively to be a desert, Ruddy Shelduck were seen and Tristram obtained a Redshank and a pair of Little Stint, the only ones seen on the whole expedition. Several Asiatic and Kentish "Dotterel" were observed and six specimens of the Ash-coloured Martin, now the Pale Crag Martin *Hirundo obsolete*, collected.

During the morning the expedition was joined by a number of armed Arabs, forty-eight footmen and fifteen mounted spearmen, but it turned out that they were known to the guards of the expedition which was now accompanied by a total of seventy-six armed men—a veritable army. Leaving the Sebkah they advanced into the Ghor, "a wild thicket and oasis of trees of various kinds, with fertile glades". "The place positively swarmed with birds in countless myriads, rising at every step with the indifference of strangership. There were doves by the score, on every bush, large and small (*Turtur risorius* and *T. aegyptius*), bulbuls, hopping thrush, shrikes, the gorgeous little sunbird, resplendent in the light, and, once more, our new sparrow. The Abyssinian larks, pipits and wagtails luxuriated in the rills at our feet." In the meantime the chieftain who had joined them had taken six prisoners, armed men who had been spotted spying on them. As a result, much to Tristram's annoyance, he would not permit a shot to be fired in case it attracted an enemy.

This decision turned out to be wise! The guard ahead discovered that much of the village at which they were to camp had been burned to the ground and was still burning; bodies lay amongst the wreckage. "Thus prematurely our hopes of the richest ornithological harvest in the country were foiled." Tristram took a small party north and, not to miss a chance, collected some sunbirds and a single male Dead-Sea Sparrow, which fell into a nest in the tree from which it was shot—"and delayed us long in retrieving it. We also secured doves and abundance of partridges for dinner." The members of the expedition were divided about continuing along the east bank of the lake, both Tristram and Bowman being in favour, as they had already paid the guards for this purpose. However, common sense prevailed and though very disappointed they decided to retrace their steps and then move westward to Beersheba from the bottom of the lake and thence north to Hebron, Bethlehem and Jerusalem.

In the morning they went back towards the previous day's carnage where ravens, kites and vultures had already descended onto the village. "Wherever the body is, there the vultures will be gathered together" (Matthew 24:28, RSV)—and the Ravens also, wrote Tristram. "Against them we perpetrated a regular *battau* on their way to their uncleanly feast . . . We brought down more specimens than we could carry away . . . Had we not been compelled to leave we might doubtless have stood among the trees, and, with the human bait before us, have continued our warfare throughout the day."

They collected specimens of all three species of ravens, but the kites and vultures took wing and rose too high for the guns. As the column departed "all the kites and vultures of north Arabia seemed to be rushing to the banquet" and the plans to explore east of the lake were, like the Sea, dead. "Encumbered with human prisoners, for our men had captured more than a dozen, and knowing that the enemy was lurking in the woods, of which the whole district was full, the keenest ornithologist might be excused if he reserved his second barrel for a bullet and declined not to wander far from camp." (Tristram, 1866) On 1 February they made their way up the Wadi Zuewirah, where at a height of 2,000 feet above the Dead Sea they were able to look back over a complete panorama of the Lisan and beyond, to which Tristram had every intention of returning in better times.

— 9 —

# The Long Winters of Palestine 2: the Retreat from Beersheba to Beirut

HOOPOE. (*Upupa epops.*)

Turning their backs on the barren peninsula and heading towards Beersheba, the expedition disconsolately climbed the track until their retreat from the Dead Sea found them on the edge of the upland wilderness of the Negev (Tristram, 1865c). During the morning they saw little apart from a single chat and two Desert Larks but in the afternoon they encountered "a long shallow basin of tender and fresh verdure, a cheering contrast with the scant vegetation of the highlands of our morning's walk." Birds abounded and Tristram forgot his earlier disappointment as he commented that "a Scottish moor could not be better stocked with game". Thirteen brace of fat Dotterel (*C. morinellus*) a rare Sandgrouse—*Pterocles guttatus*—and many larks of the "Saharan group" were obtained. As the first of these (boiled and grilled) were being consumed they were visited by a single wolf, and Tristram slipped a bullet down the barrel. It appeared much larger than the European Wolf and much lighter in colour. A carefully placed shot between its front feet precipitated a slow retreat.

After camping overnight, during which there was a hoar frost, food was running low. Offered a sheep by a local inhabitant Tristram declined the offer because of the expense, and instead shot twenty brace of Dotterel. Seven species of larks and a few Bush Chat (*Saxicola philothamni*), a species he never encountered elsewhere in Palestine, completed the morning bag. Moving on, the party encountered many Common Cranes and sandgrouse and towards the end of the day they came upon the roosting ground of the cranes. This was on a gently sloping knoll, marked like the roost of sea fowl, "where no ambush was possible and where a good look-out could be kept on all sides. Their whooping and trumpeting enlivened the watches of the night and all night long we could hear flocks passing overhead on their way to their quarters close by."

The third of February dawned very cold, and early in the progress north Upcher brought down several Spotted Sandgrouse and an Asiatic Plover—the Caspian Plover *Charadrius asiaticus*. For a time the latter species was plentiful during their progress which was carried out on foot—the horses having been sent on ahead—and cranes continued to pass over. A few Houbara Bustards *Chlamydotis undulate* were seen as they approached Beersheba. After an overnight camp Tristram rose early. He put up an Eagle Owl *Otus ascalaphus* and saw eagles and cranes in the distance. Later, pausing for the mid-day meal of eggs, barley-cake and grilled plover, the arrival of several thousand Arabs, from all sides, predicted trouble, and they accepted advice to move on in case they found themselves in the middle of a battlefield. They were effectively driven north-east and prevented from visiting Kadesh Barnea or penetrating to the coast near Gerar and Gaza, as they had intended. However, they were thankful to have avoided being caught in the impending fight between the Arabs and Turks, and before dusk life was back to normal, with observations of eagles and the shooting of a specimen of Audouin's Gull and bags of plovers and dotterels. Tristram observed that they seemed to be destined to be in the midst of Arab frays but he seems not to have been particularly put out.

From fleeing the battle zone they had been journeying over rolling hills and green plains, from the Negev to the hill country of Judah, where gulls were not rare and apparently fed on snails which covered the plants. By sunset they arrived near Tell Hhora and "a large cave full of sweet water". The walls of ruined buildings of the old Israelite city near their camp sheltered both owls and hundreds of Rock Doves and after a second night they moved on northwards. Passing from the southern countryside of Judah into more hill country, with changing flora and fauna, the desert larks gave way to Crested Larks and Sylarks, the Sandgrouse to Grey Partridge and the Dotterel and Caspian Plover to Lapwings; no cranes flew overhead. Bartlett shot the first Red-tailed (Long-legged) Buzzard *Buteo ferox* of the expedition, described by Tristram as "a rare and magnificent eastern species" which was later to be found more commonly, and Upcher brought down a Red Kite. Continuing towards Hebron through the Vale of Eshcol, darkness and the rain came down. Bartlett had earlier become separated from the main party but fortunately encountered them before they saw the lights of the city.

Next day Tristram explored Hebron and on returning, again after dark, found that "many birds had been collected, all of which were the same as those of Carmel and Mount Ephraim—jays, woodpeckers, owls and finches—telling us we had got back to the central country and need expect no more of the rarities which had rewarded us in the south." Next day, after leaving the city, they headed north with Solomon's Pools as the objective. Passing out of the Vale, English birds seemed to be the only inhabitants of the hillsides, where, apart from the Greek Partridge, Goldfinches, Buntings, Woodlarks and Linnets were the only species observed. Eventually the group came upon the Pools, three vast man-made reservoirs on which flocks of wild duck floated. These were mainly Gadwall, Pochard and Shoveler (Plate 45) and were so disturbed by the arrival of the large group that only a single Pochard succumbed to their guns. However, Tristram commented that he "received in Jerusalem a single Wild (Whooper) Swan (*Cygnus musicus*), now *Cygnus cygnus*. Shirihai (1996) reports this as the first record of the species in Israel, shot on 23 December 1863.

The party passed through Bethlehem, moving on straight away to Jerusalem. There they spent ten days packing the equipment and making repairs, before heading off to the coast and Jaffa. Here Medlycott was to leave them, but on arriving at the port they found that the steamer had sailed. However, the expedition continued on its way and before leaving the environs of Jaffa Tristram obtained a fine Peregrine Falcon, the first to be collected on this visit to Palestine. On the plain outside the town larks abounded and four species were seen—Calandra, Sky, Crested and Wood—together with quails, buntings, starlings and Sardinian Black Starlings. Running round the watery lagoons were Ringed and Kentish Plovers and on the more distant lagoons herons, Squacco Herons and egrets were feeding. Heading north for Nablus the ground was carpeted with spring flowers and the olive groves swarmed with jays, owls and woodpeckers. Reaching the Merj el Ghuruk—the meadow of sinking—the party encountered a small lake on which several Black-tailed Godwits fed and Shepherd shot a Marsh Harrier. In his account of the area, Tristram (1865c) comments: "On revisiting the spot in April I found the water still remaining, and the stilt and other species of waders, as the 'zic-zac' or spur-winged plover [Plate 46] and the little ringed (plover) dotterel, breeding in the marsh."

On 26 February, they reached the Sea of Galilee and approached it from behind the old town of Tiberias where the view prompted Tristram to quote from Byron's 'Destruction of Sennacherib'. The following morning "the sunrise was as majestic as the moon had been lovely" and they set off for an excursion to the plains of Gennesaret. Tristram records that "in every way we were repaid for our excursion. Scenery, fish, birds, butterflies, flowers, shells—in all we gathered a harvest. U. [Upcher] bore home a Bonelli's Eagle in triumph and we secured several grebes and gulls having to be our own retrievers, and to take no less than three swims in the lake to fetch out our game."

Later on in the day they visited the eyrie of a Cinereous (Black) Vulture *Aegyptius monachus* which had been described to them. It was situated at Ain el Barideh and was located in a cave to which they climbed. They were ten feet below the cave when the bird flew out and it was the first of the species that they had seen on the expedition. Shepherd climbed to the nest and collected the single large egg which was the "first oological capture of the season". The bird circled outside the cave for some time providing good views so that the identification was made certain.

On 29 February Tristram took a sailing boat on the lake and obtained "two or three Great-crested Grebes and a single Royal Eagle Gull", now the Great Black-headed Gull, which Tristram described as "by far the most magnificent species of its kind in the world". After a very uncomfortable journey back to Tiberias, in which the boat was buffeted by the rain and high winds, they made the shore. Tristram insisted on continuing his journey to Gennesaret, to where the expedition had moved during the day. Of this journey, accompanied by two guards, he reported that "heavily laden with my burden of gulls and grebes, I had a weary walk over the rocky ground, in the dark, and when we reached the plain missed the path to our tents, which we did not recover until our signal guns were heard and answered." Yet again, this was an indication of Tristram's adventurous nature—perhaps over-adventurous—and his fitness too.

The period between 1 and 8 March was spent exploring the western coast of Galilee where their camp was situated 500 feet (152 m) above the level of the "sea" and 800 feet (244 m) in front of "beetling cliffs" which housed scores of Griffon Vultures, Lanner Falcons and Ravens. On the lake the waters were "of a crystal-like calmness mirroring the great sea-birds, eagle-gulls and cormorants, which lazily flap their heavy wings over it." Each morning Tristram and his colleagues took a dip in the Basin of Ain Mudawarah, a man-made, enclosed water about 30 yards (27 m) across surrounded by eight-feet high walls within which the water was some three feet deep and fed by a central fountain. "Several wild fig trees hung in fantastic shapes over the sides of the bath, and slender oleander bowed their pink tufts of blossom to the breeze, whilst the gorgeous blue and red kingfisher *Halcyon smyrnensis,* sat motionless, watching for its prey, and francolins and quail called incessantly in the marsh and bean fields."

Later in the month (31 March) Tristram, together with a mounted guide, took an excursion from this spot to the ruins of Khan Miniyeh, to complete his survey of the western shore and here he found the nests and eggs of the Common Kestrel and also those of the Red-rumped Swallow. "Closer to the shore was a luxuriant *Papyrus* marsh in which the stems measured up to sixteen feet in length, with a diameter of three inches. This thicket was the home of (besides the Smyrna Kingfisher) the Great White Egret, the Little Egret, the Bittern, the Little Bittern and the Purple Gallinule *Porphyrio hyacynthus* [now *P. porphyrio*], all of which I put up in a few minutes." Moving on they approached the upper Ghor or flat alluvial plain, where the Jordan flows into the lake. White Storks, herons, Spur-winged Plover and Gull-billed Tern *Gelochelidon nilotica* were common and Tristram shot a Great-crested Grebe in full plumage.

Northeast of the lakeside, along the very dark, narrow gorge of the Wadi Leimun where the party next ventured, they encountered limestone cliffs, 500–700 feet (152–213 m) in height, scarred by caves into which the sun never penetrated. These were the nesting places of the Griffon Vulture, Lammergeyer, Lanner Falcon, several species of eagles and myriads of Rock Doves *Columba livia*. Tristram gives a very vivid account of the last species: "In absolute clouds they dashed to and fro in the ravine, whirling round with a rush and a whirr that could be felt like a rush of wind. It was amusing to watch them upset the dignity and the equilibrium of the majestic griffon as they swept past him. The enormous bird, quietly sailing alone, was quite turned on his back by the enormous rush of wings and wind." An isolated rock stack housed Griffon Vultures on all sides and a Wall-creeper showed itself on the rock face high and out of range, and above Alpine and Galilean Swifts screamed through the air. Whilst this was obviously a very enjoyable day they had little to show for it in terms of collecting. However, they had much more success in the Wady Hamam, at the south west end of the plain, where they spent three days exploring. Here the cliffs reached about 1,500 feet (457 m) in height but rose in steps of broken rocks from the bed of the wadi which itself was half a mile in width. Giacomo, a dragoman/guide, was an expert rope climber and obtained "a good harvest of griffon's eggs". "Five great griffons were shot by S. [Shepherd] and U. [Upcher], the preparation of whose highly scented skins was no light task for the taxidermists." In the three days in the wadi the climbers obtained fourteen nests of griffons but no Lammergeyers as the young had already flown. Tristram was particularly pleased that Shepherd and Upcher obtained several specimens of the Galilean Swift and these were the first to be returned to Europe.

On 9 March they struck camp and headed for Tiberias but were stopped on the way by the arrival of the monthly mail. A letter from their contact Zimmer informed them that they were now able to move east of the Jordan so they headed for Nazareth to make the necessary arrangements. The time had now come for Shepherd and Upcher to leave the expedition and return to Europe so that the most energetic collectors were lost to them.

In Nazareth, the dragoman Giacomo announced that he would not go east of Jordan, as he did not want his throat cut, but Tristram was adamant and surprised their guide/interpreter by saying that he might remain in Tiberias and await their return. Again Tristram was not going to be easily discouraged from following the planned route, though others might have thought it a rash decision. The depleted group passed through the Wadi Bireh and crossed the river south of Galilee by the ancient bridge, then turning along the left bank of the river, putting up herons, Spur-winged Plover and quails continually with Great Black-headed Gulls and vultures overhead. Passing through the Ghor they began the climb to the forested plateau, clouds of Wood Pigeons breaking from the evergreen oaks, and jays and woodpeckers in every glade.

By 14 March it became evident that Giacomo's reluctance to cross the Jordan had been well founded. First the expedition had found itself in the middle of a running battle between a party of goat-stealers and the owners of the goats and then on arriving in the village of Suf it became evident that the cavalcade would not be allowed to leave without excessive payment. Guards were disposed for the night's camp but as this was struck on the morning of 15 March they were surrounded by approximately a hundred and fifty armed men and boys. Tristram had to forcibly disarm one of their own guards to prevent a battle, but with some clever manoeuvring of the donkeys managed to get his armed group between the donkeys and the armed villagers. Then, by scattering money they got onto open ground. Their original plan, as they had given up the idea of progressing to Gerash, was to return to Tibneh, but this proved impossible as their retreat was cut off. Now at a height of 3,500 feet (1,067 m), above the plain of the Ghor, the party decided to cut off westwards, leaving Jebel Ajlun to the left. Late on 15 March they reached Tibneh and friendly faces.

Rising early the following morning they collected a pair of Great-spotted Cuckoos, a new chat and other birds, amongst them partridges for dinner. They then rode for the bridge they had so recently crossed in the opposite direction. A camp was established at Caiffa, Bowman was put on the boat home and the next ten days were devoted to activities around Mount Carmel. "The birds were not many in kind except the great birds of prey." Birds collected included vultures and eagles of all sorts, wood pigeons, black-headed jays and shrikes of three species—Great Grey Shrike (*Lanius excubitor*), Masked Shrike (*L. nubicus*) and Woodchat Shrike (*L. senator*)—Pallid Harrier (Plate 47) and two sunbirds.

On 27 March Tristram's party accidentally came upon a recently arrived group of Englishmen, Messrs. Egerton-Warburton, Cochrane, Barneby and Bateman, and the two groups decided to travel together. Leaving Caiffa they proceeded westward towards Nazareth and the Wadi Bireh. North of Esdraelon they added the Honey Buzzard and many summer birds to their list, and then in Nazareth called in on the Turkish Governor. Tristram was still anxious to venture east of the Jordan but the Governor gave him no encouragement and the combined groups continued to Tiberias. There the lake was three feet shallower than when they had last seen it and the Great Black-headed Gulls had gone but the grebes were still present—at least those which Tristram had not previously "collected". On Easter Day the White Storks (Plate 48) appeared and as Tristram pointed out, the stork "knoweth her appointed times" (Jeremiah 8:7). This was the first time that they had seen the birds passing over northwards in their thousands and on the next day the plain of Gennesaret was covered with them in every direction; two days later they were gone. Another sign of spring was the sighting of the "beautiful russet swallow—the Red-rumped Swallow—skimming over the lake and plain till sunset".

They camped well above the level of the lake for four days, "high enough to avoid the malaria of the lake", and continued their collecting. Tristram regarded their greatest trophy here as the nest and eggs of the Galilean Swift—now the White-rumped Swift. They discovered the nest of the Palestine Sunbird and climbed to several vultures' eyries. Savi's Warbler *Lusciniopsis* [now *Locustella*] *luscinioides*, River Warbler *Locustella fluviatilis* and Cetti's Warbler *Cettia serricea* [now *cetti*] and many other rare species skulked by the sides of the stream and amongst the papyrus, but, though heard continually, were most difficult to obtain.

At this stage the two parties were thinking in terms of pushing south as far as Jericho, but initially followed the south-westerly track towards the western end of the Wadi Bireh. Tristram regarded the first part of the journey as boring, covered by wide expanses of barley, but was cheered somewhat on encountering the wooded areas just east of Tabor where the spring migrants were becoming established for the breeding season. The European Roller *Coracias garrulus* and European Bee-eater *Merops apiaster* were both collected here. Now there was more bad news for Tristram as the plan to descend the Ghor was considered impracticable because of the presence of "several tribes of evil repute encamped below Beisan". After some consideration Tristram decided to leave Lowne and Bartlett with the greater part of his original expedition to work the region of the Tabor. He would then proceed with his new friends southward initially, in the hope of further exploring the eastern regions. Certainly Tristram was not one to give up on his original plans of crossing the Jordan and this was obviously a good opportunity to do so. However, in taking only the smaller tent, Hadj Khadour and the boy Elias, two horses, a mule and an ass, this was perhaps not the best idea he had ever had.

From Beisan Tristram and Egerton-Warburton's party proceeded apace westwards to Jenin then south to Jerusalem, a journey of about eighty miles, which they accomplished on 9 April, in just over three days. Not surprisingly Tristram recorded little on the way. On the first day of the journey when they stopped around midday for refreshment, Tristram records that "a Black Kite came down to share our meal, which we shot, as also the Ortolan Bunting, being the first of either of these migrants that we had seen." From where they had stopped they could look across the Ghor and beyond the Jordan and it was clear from his writing that this area was still very much on his mind. Parties of White Storks, obviously exhausted from their migratory flight, remained close to the track as the expedition passed and on 9 April they encountered small flocks of Rock Thrush spread out over the hills. Then they were in Jerusalem from where Tristram was to replan their journeying! This became even more important as it was discovered that their old friends the Jehalin had been set upon by bandits, who had killed fifteen and injured thirty-eight of their number. The bandits were now occupying the area east of Hebron which was the objective of Tristram's new fellow travellers, making it impossible to proceed there.

Remarkably Tristram decided to embark on a ten day sojourn to Jericho in order to compare the summer flora and fauna of the Ghor with that of the winter — "alone" as he put it but with three Palestinian helpers. That this was more than a little foolhardy was demonstrated by their being approached after setting up camp on their first night out from Jerusalem by a group of four self-described "robbers" on their way to Jerusalem with thirteen stolen camels. By sleight of hand, assisted by his friend Jameel, he managed to make his armoury appear more extensive than it was by reloading his single revolver four times! The "robbers" departed, with their camels but without event, and once again Tristram had been very lucky.

The following week (12–19 April) Tristram spent collecting in the "ciccar" of Jordan. The nests of the Bulbul, Sunbird, Fantail and *Crateropus* "and many others" were found and he observed that the birds of the Ghor were very similar to those of the surrounding country. Bulbuls and Nightingales were in full song and the "Turtle Dove, just returned . . . stocked every tree and thicket." Houbara Bustards were common on the barer parts of the plain though Tristram searched in vain for a nest and had no more success with the Sandgrouse; clearly it was too early in the season. On 18 April the rest of the expedition joined Tristram and they decided to explore the gorge of the Wadi Kelt. Here they took the eggs of two pairs of Ravens and watched a flock of Rollers excavating a soft bank. On the next day they rode to the head of the Dead Sea, as far as Ain Feshkhah. Here Tristram had his "opera glass" stolen—"an irreparable loss in bird-nesting". The rest of the group left for Marsaba and Tristram returned to his lonely tent.

On 20 April Tristram returned to Jerusalem and spent several days dealing with matters other than collecting. On 27 April he set off towards the Jordan intent on eventually making the crossing. Reaching its banks he found the trees full of doves and singing nightingales. After an exciting crossing of the Jordan they reached Keferein where they found the little Hey's Partridge to be numerous. Many of the young ones were run down on foot for the pot. The Greek Partridge was equally numerous despite the fact that the raptors flourished, "not extirpated as they were in more civilised climes". In his description of the area Tristram commented: "Some of our guards, having discovered our 'fantasia' for eggs, searched with success for nests, and altogether we agreed that we had fallen amongst a by no means disagreeable set of savages." Passing the castle of Hyrcanus the expedition encountered a row of caves off a ledge above the track. Tristram set about exploring these and was pleased with the results: "We had a most successful natural history exploration in these caverns, having taken amongst us the nests of two vultures, the largest Egyptian Owl, lesser kestrel, and our first nest of the russet swallow, besides the bulbul's in the castle."

By the last day of April they were again travelling north towards Amman which would be the most easterly point of their wanderings, a word which accurately describes the route that the expedition had taken since its beginnings in Beirut. Overnight they camped in the Wadi Heshban and in a nearby ravine they observed the nest of an Egyptian vulture, in an old hermit's cave. One of the boys climbed to it and collected the eggs. Approaching Amman itself, news of the expedition had obviously preceded them as awaiting was a group of Bedouin with quantities of eggs of hawks and vultures for sale. Such were the channels of information that the Bedouin already knew the prices which Tristram was prepared to pay for these trophies, and they were advised to follow the expedition to the campsite within the bounds of the old Roman Philadelphia to negotiate the sales. During the previous three days Tristram recorded that they "had reaped an amazing harvest of eggs ... especially at Heshbon and here vultures, eagles, Great-spotted Cuckoos and some dozen of other species were collected. Our scouts found the nests and pointing them out to us as we rode, many a box was filled."

Waking on the morning of 2 May Tristram found that "the ruins swarmed with Jackdaws, not the race (*Corvus collaris*, Drummond) which inhabits the Ghor, but the common Jackdaw of England, the same species we had taken on Mount Gerazim, and the Great-spotted Cuckoo (*Clamator glandarius*. L.) had been depositing its eggs in the nest of the Hooded Crow." On the slopes of Mount Gilead, Cochrane and Tristram discovered a cave and whilst collecting ammonites in it disturbed two Alpine Swifts. Searching they found the nest in a crack in the rock. Whilst they were able to touch the eggs they could not remove them from the nest and had to leave them, much to their annoyance, as they were something of a rarity. During the next few days Tristram recorded little of ornithological interest other than the offer by boys from one of the Gilead villages to provide the eggs of woodpeckers and rollers. Whether or not they did so is apparently unrecorded. The main object at this time was to reach and cross the Jordan whilst avoiding problems with the local inhabitants, and, on 6 May, just before arriving at the village of Arad on the fringes of the river, they encountered myriads of Turtle Doves, mainly Common Turtle Doves, which they put up in scores from nearly every bush. Of more interest, however, were the Marsh (Spanish) Sparrows which were nesting so commonly as to bear down the branches of the trees with their weight, and Tristram commented on the birds being "literally deafening" and it being "scarcely conceivable how such multitudes can be fed, but the bushes and weeds were laden with berries and seeds." As they recrossed the Jordan he felt that he had re-entered civilization.

After travelling west to Nazareth, Tristram and his friends Egerton-Warburton and Cochrane met up with Lowne and Bartlett in Banias with the plan to explore the area west of Galilee during the next three weeks. Hadj their muleteer had heard them discussing the fact that the ducks, grebes and gulls were absent from the lake and one morning excitedly announced that he knew all about them and that his friends were coming with a hundred grebes' eggs. On the plain of Gennesaret they met up with a group of Arabs carrying a large basket and on examining its contents discovered that it contained a quantity of fresh-water mussels. They insisted that these were the eggs of grebes and this belief seemed to be universal. Tristram commented that the *Norwegian* fable of the Barnacle Goose was here almost exactly repeated; apparently he had not heard of it in England, where the swan barnacles on Peel Island, near Barrow-in-Furness, were once thought to give rise to the Barnacle Goose *Branta leucopsis*.

North of the Sea of Galilee is the Lake Huleh where the expedition, now reconstituted, found large numbers of warblers and Red-backed Shrikes breeding on the southern slopes: "in three or four hours we obtained about twenty nests, chiefly the Orphean Warbler and Lesser Whitethroat as well as Cretzschmar's Bunting." Further along the shore they "saw many herons, grey, purple, white, buff-backed (Cattle Egrets) and squacco, and shot a number of pratincoles, as well as both species of cuckoo and the bright golden oriole." Three miles north of Lake Huleh they came upon the edge of the great marsh where all the species of herons that they had seen on the lake were present in abundance, together with Little and Great Bitterns, Purple Gallinule, Marbled Duck *Mamaronetta angustirostris* by the hundreds "and whatever else loves a jungle and a swamp, with frogs for dinner". Marked on the maps as impenetrable, the marsh proved to be so. They gave up trying, "satisfied that the marsh birds were not to be had". However, they shot two Bitterns later and in attempting to retrieve one of these Tristram fell in, saved only by his gun falling across *Papyrus* stems; he certainly lived a charmed life.

Moving towards the bridge of Burghuz, the gorge narrowed and was some one thousand feet deep (305 m). Looking over the edge they could see the opposite side perforated by many caves, the eyries of vultures, eagles and Lanner Falcons. Large numbers of these circled above them as they made their way down to the bridge and up the other side towards the village of Burghuz. This was the cleft of Leontes, the only true alpine scenery they had encountered throughout the expedition, and from here they moved on to the little town of Hasbeiya. Leaving this and approaching El Kuweh they came upon Blue Rock Thrushes in a hidden chasm with Rock Doves and circling Crag Martins. Above, where the chasm widened and banks of green turf clung on in areas of loose rock, Wallcreepers were present together with what Tristram considered to be two separate species of Nuthatch, the Syrian Nuthatch *Sitta neumayer* and the European Nuthatch *Sitta europaea* (Plate 49), both flitting in small parties from side to side of the gorge. The latter is now a rarity in Israel. The former Tristram collected in the Gorge of the Leontes and this is recorded in his Catalogue. His *Fauna and Flora of Palestine* (Tristram, 1884) records a third species *S. krueperi* from Leontes. Sclater (1865) points out that a specimen, shot in Hermon by Bartlett and claimed to be this species (Tristram, 1864b), is in fact *S. europaea* and Tristram (1866) accepts this. However, in the same paper Tristram writes that he "possessed two of Dr Kruper's type specimens and felt confident that I frequently saw this little nuthatch in the Leontes gorge. I shot them but was unable to recover the specimens in that tremendous depth but saw them closely enough to identify the chestnut collar; and Mr Cochrane took a nest in this place which he kindly shared with Tristram, the eggs of which are only half the size of those of our Common Nuthatch, and doubtless belonging to this species." The eggs of *S. krueperi* are significantly smaller than those of *S. europaea* which in turn are very much smaller than *S. neumayer* (Table 6). It is unlikely that Tristram would have again wrongly identified Kruper's Nuthatch after his experience with the Hermon birds, particularly since it is so distinctive and uniquely sports an orange-banded breast. In addition, by the time that he published his *Fauna and Flora* in 1884 he possessed two of the type specimens of this species which he received from Kruper himself, taken in Smyrna on 9 September 1864. It is thus possible that this species has been omitted from the Israeli list because of confusion with the birds collected in Hermon, and it is still to be found breeding less than 150 miles to the north.

Passing mount Hermon in the sunset the expedition encountered the town of Rasheiya "perched on a spur of Hermon projecting to the north; the palace of the Dreuse Emir, the hereditary feudal lord occupying the brow and the straggling flat-roofed houses, the slopes and depressions on the irregular site bearing a rude resemblance to the city of Durham." Probably even the adventurous Tristram was thinking of home!

They made Rasheiya their headquarters for some five days and here, amongst the vineyards they found three birds new to the expedition, which Tristram considered at the time to be new species. The first, a small finch, he named *Serinus aurifrons* which became Tristram's Serin (Plate 50), but later was found to be conspecific with *S. syriacus* now usually referred to as the Syrian Serin though in some circles it remains as Tristram's Serin. The second was a small warbler that Tristram named after his fellow explorer *Hippolais upcheri,* Upcher's Warbler. Again, this had been previously described and thus became *H. languida* but has retained its English name of Upcher's Warbler. The third Tristram named *Bessanornis albigularis* the White-throated Robin Chat. Here both names had been preceded so that it is now The White-throated Robin *Irania gutteralis.* With communications as they were in those days it must have been very difficult to decide if something was new, particularly during an expedition cut off from both colleagues and books, and disappointing when it became evident that your discoveries had been made previously.

All three of these were found over a period of two days and the nests and eggs of all three species were acquired. Much cheered by their finds they decided to go to the summit of Hermon, which though steep they did not find too arduous as they were not forced to walk and remained mounted. Clearly Tristram was very much enjoying Hermon and they reached the top after a five hours climb—"turfy banks, gravelly slopes and broad snow patches alternated till we reached the summit. . . . To us they were rich indeed." Small groups of Alpine Chough hovered around them, and ravens and Common Swifts croaked and screamed above. Both Griffon and Egyptian vultures and an eagle or two circled above and several species of smaller birds were living there and apparently breeding abundantly. The Linnet *Linota (Carduelis) cannabina,* Common Wheatear *Saxicola (Oenanthe) oenanthe* and Snow Finch *Montifringilla nivalis* all had young in their nests but for the Persian Horned Lark *Otocoris pencillata,* now regarded as a form of the Shore Lark *Eremophila alpestris.* Tristram records that "we obtained only one sitting on eggs" and he was particularly attracted to these birds: "It was a beautiful sight to watch these larks scattered over the dome of Hermon, warbling their rich yet subdued notes, with erected crest, on the desolate tops of the rocks which strew the summit." This population of the Shore Lark is now recognised as forming a "semi-endemic" race *E. a. bicornis* on Mount Hermon. It is recognised from other forms by its white throat and greater extent of black on the breast, and Tristram comments: "Of the beautiful Horned Larks, Palestine affords the finest" (Tristram, 1866), and further: "It was a strange surprise to discover on this arctic patch two English winter birds, with the horned lark of Persia, the chough of the Alps, and just below, a finch related to the Himalayan birds (Syrian Serin), and a warbler related to the central African *Bessornis* (White-throated Robin)." (Tristram, 1865b)

On 6 June the party left Rasheiya for Damascus, departing from Israel as we now know it, and for some days their interests lay elsewhere than ornithology; on 11 June they set up a two-night camp at Ain Fijeh and bathing in the stream there found themselves sharing it with the little water ouzel—the Dipper of Tristram's Northumbrian boyhood. "What a bird contrast!" he writes, "A few hours earlier we had shot the African Sand Grouse; here we were watching the ouzel of Northumberland." Clearly, towards the end of a long expedition, his thoughts were turning to home. By 13 June they were well into Syria and above Surghaya, on the watershed of the Mediterranean. Here they encountered the Rock Sparrow *Petronia petronia* which was very common on the open ground as was the much rarer Pale Rock Sparrow *P. brachydactyla*, two nests of which they obtained. The Hobby *Falco subuteo* and Eleonora's Falcon *Falco eleonorae* were frequently overhead and Cetti's Warbler sang from the scrub.

Now heading westward towards the Lebanon, on 15 June they obtained many rare birds including what Tristram recognised as the Syrian Redstart *Ruticilla semirufa*, which he claimed was unknown in English collections. Whilst this lays dark blue eggs, which Tristram took in Palestine, the Black Redstart *Phoenicurus ochruros*, of which it is now considered conspecific, lays white eggs, and this may have gone some way to confuse Tristram.

Much of the snow had gone from the high ground and the party was able to cross by the highest pass close to the summit of Lebanon where the snow was hard and compact, then descending onto the cedars. Here the birds and plants were similar to those on Hermon, and going down the western slopes they again encountered the Linnets, Wheatears and Shore Larks. Little flocks of Coal Tit *Parus ater* and a few of the Russian Sombre Tit *Parus lugubris* were to be seen amongst the trees. Neither of these species had been obtained previously in Syria, according to Tristram, and they were then still in that country, the borders of the Lebanon not having been extended until after the First World War. Tristram's "new" Serin and Black Redstarts sang from the trees as they made their way to Hazrun where they were to camp for the night. Their servants had arrived before them and had announced their taste in natural history so that on their arrival they were greeted by a crowd of urchins with trophies to sell. There were squirrels, birds and a score of nests "to give us employment after a day of fourteen hours' exciting travel".

From Hazrun to Akurah and from Akurah to Meiruba they spent a long time admiring the cedars, to the benefit of the wildlife, except for the pair of Alpine Choughs shot on the latter of these two journeys. On the morning of 21 June Tristram "mounted early and accompanied by Mamoud, my faithful henchman, left my sleeping companions to follow at their leisure. At 10 o'clock we dismounted at the door of Constantino's Hotel, in Beyrut, and my wanderings in the Holy Land were ended." . . . at least for the time being, but return he would.

Table 6: Dimensions of the eggs of three species of *Sitta*. Length × Breadth in mm.

|  | *S. europaea* | *S. krueperi* | *S. neumayer* |
|---|---|---|---|
| Makatsch (1976) | 19.20 × 14.32 | 17.33 × 13.29 | 20.60 × 15.25 |
| Schonwetter (1984) | 19.4 × 14.3 | 17.0 × 13.0 | 21.2 × 15.2 |
| S'wetter Range | 16.5–22 × 13.2–16.0 | 15.9–18.4 × 12.6–14.0 | 18.5–23.4 × 14.0–16.5 |

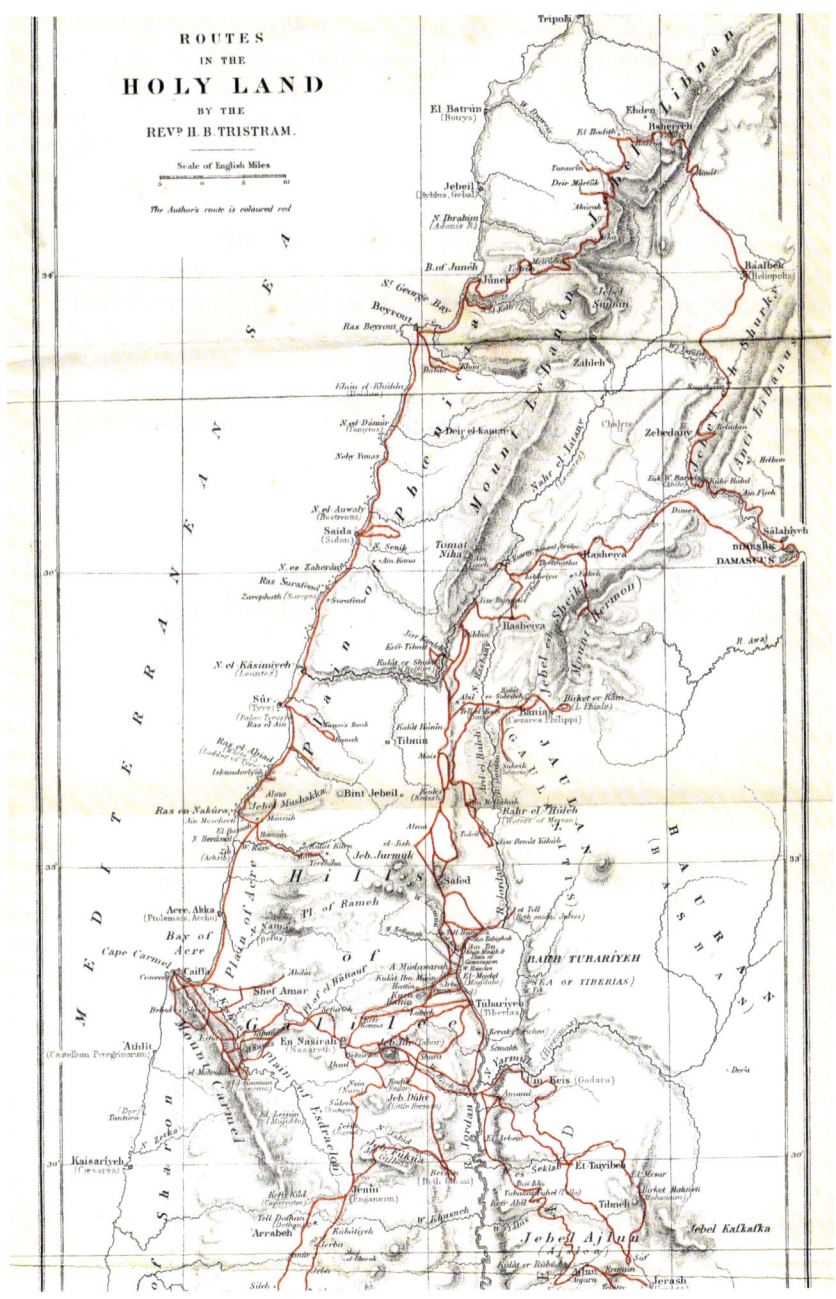

Figure 5: Map of the northern part of Palestine showing the routes taken by Tristram during his expeditions. From H. B. Tristram, *The Land of Israel*, 3rd rev. edn., 1876.

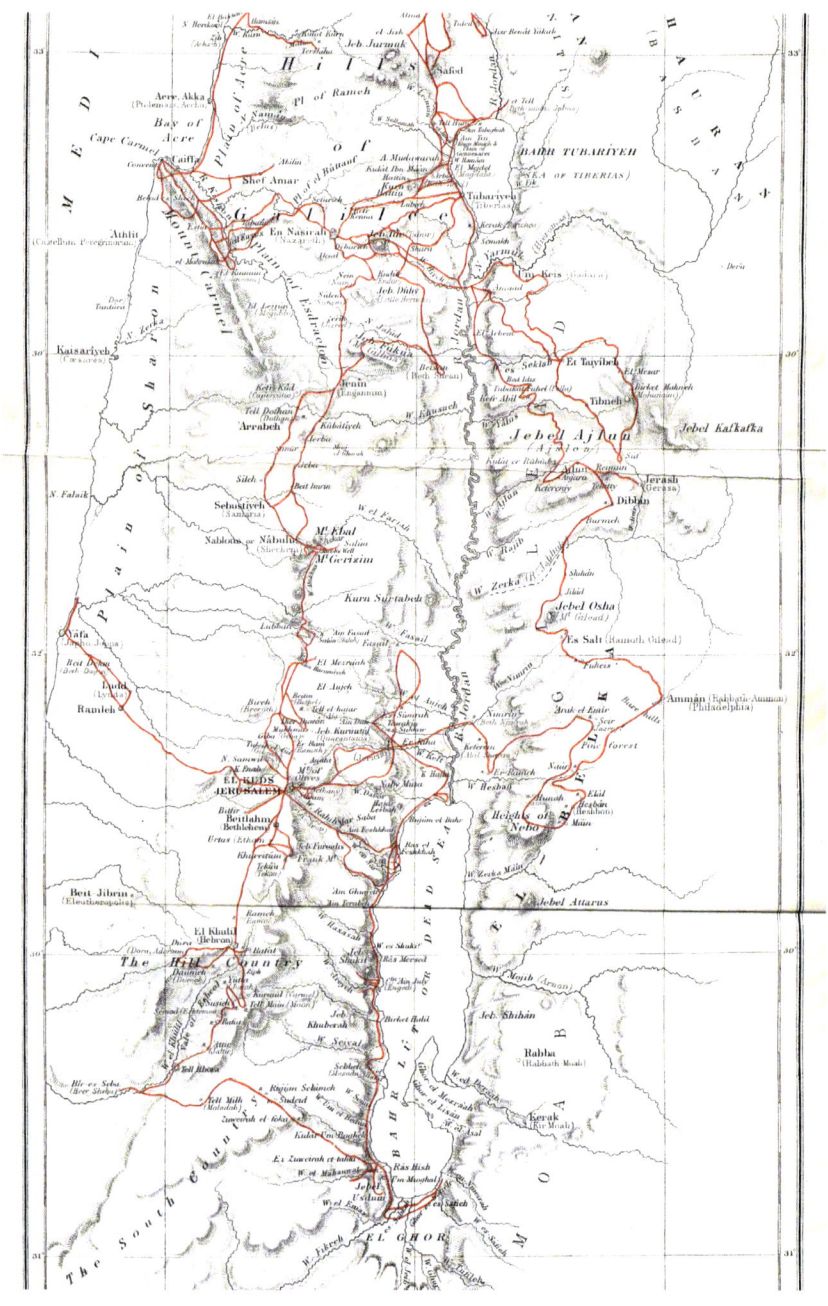

Figure 6: Map of the southern part of Palestine showing the routes taken by Tristram during his expeditions. From H. B. Tristram, *The Land of Israel*, 3rd rev. edn., 1876.

Plate 45: Shoveler *Anas clypeata*. Common passage migrant and winter visitor with some birds summering. Known to have bred once.

Plate 46: Spur-winged Plover *Haplopterus spinosus*. A common resident breeder, widespread in winter.

Plate 47: Pallid Harrier *Circus macrourus*. A common passage migrant and occasional winter visitor.

Plate 48: White Stork *Ciconia ciconia*. As in Tristram's time this is a common passage migrant, commoner in autumn than in spring and occasionally stays to breed in the north.

Plate 49: Nuthatch *Sitta europea*, courtship feeding. A breeding resident around Galilee and in the Lebanon in Tristram's time; it is now very rare.

Plate 50: Syrian Serin *Serinus syriacus* (above) is Tristram's Serin *S. canonicus* now regarded as conspecific with *S. syriacus*. Dead Sea Sparrow *Passer moabiticus* (below) – discovered and named by Tristram. From Tristram's *Fauna and Flora of Palestine* 1884.

**Plate 51: Cyprus Warbler** *Sylvia melanothorax*. This new warbler was first named by Tristram as the Black-throated Warbler after he had identified it by means of its call note from amongst Sardinian Warblers. From H. E. Dresser's *Birds of Europe*.

**Plate 52: Black-winged Stilt** *Himantopus himantopus*. Several birds were seen feeding in pools after heavy rainfall.

**Plate 53: Skylark** *Alauda arvensis*. A flock consisting of several species of lark, including skylarks, was observed feeding on caterpillars around an old water tank (reservoir).

Plate 54: Common Teal *Anas crecca*. First seen on this expedition on the Pools of Solomon; outside Jedda the species was not uncommon. A bird collected was thought by Tristram to be a cross between a Teal and Mallard (Yarrell's "Bimaculated Duck") but was almost certainly a Baikal Teal *Anas formosa*.

Plate 55: Hoopoe *Upupa epops*. A single individual sighted by Tristram was regarded by him as the first sign of spring.

— 10 —

# Palestine: The Land of Moab (1872)

DESERT PARTRIDGE. (*Ammoperdix heyi.*)

In 1871 the British Association renewed its grant of £100 for Tristram's proposed expedition to Palestine, doubled it, and formed a Committee for the Geographical Exploration of the Moab. So on 10 January 1872 the expedition left England for Jaffa where they arrived on 22 January, having called in at Brindisi and Alexandria on the way. The party consisted of C. Louis Buxton of Trinity College Cambridge; cameraman, R. C. Johnson of

Liverpool; astronomer, surveyor, mapping and second photographer, W. A. Hayne, Trinity College Cambridge; botanist, Mowbray Trotter, also of Trinity College; a cook, and Tristram. Mr Klein, Church Missionary Secretary and Representative in Jerusalem, joined them in Palestine. The main objectives of the expedition were geographical and concerned particularly with the search for old biblical sites. However, this was not to stop Tristram watching and collecting birds, though it did limit the recording of his observations (Tristram, 1865b and 1873); unattributed quotations in this and the next chapter are from these two works.

Once in Jaffa Tristram's only record consisted of his observing Teal at the Pools of Solomon. The party camped south of Hebron, having decided to cut across country from Hebron to Engedi, a route which they were reliably informed did not exist. Nothing daunted, however, they found themselves at the watershed of the Mediterranean and Dead Sea, in a true wilderness. On the way they had encountered only a few Rock Partridge (Chukar) and as the rain fell they reached the top of the pass down to Engedi, passing the cliff of Ziz and entering Engedi on 2 February.

The next day the party explored the most interesting sites of Engedi, including the caves that Tristram had visited on his previous expedition. Here, amongst many small birds making the oasis their winter quarters, he obtained four specimens which he considered to be of a new warbler, "something like the Sardinian Warbler". This was subsequently named *Sylvia melanothorax*, the Black-throated Warbler (Plate 51), but is now known as the Cyprus Warbler. Tristram commented: "My attention was first directed to them by the note, which differs markedly from that of *S. melanocephala*", the Sardinian Warbler.

Leaving Engedi the expedition turned right and headed south down a birdless and plantless shore and set up camp in Masada. The following Sunday they achieved the ascent of Masada along a quiet, zig-zag track to the cliff top, near to which, in two openings in the cliff, a pair of Lanner Falcons and a pair of owls (probably Eagle Owls) nested. On the return journey to the camp they encountered a pair of Hooded Wheatear *Oenanthe monacha* which Tristram had only seen previously on the salt-mountain of Jebel Uslum. That night they were entertained by a magnificent aurora borealis and Tristram recorded that he had never seen one so brilliant.

Rising early the next morning they continued to move south along the western side of the Dead Sea and for two days experienced a park-like wilderness of trees, shrubs and bare clearings populated only by the large Indian Turtle Dove (now the Collared Dove *Streptopelia decaocto*), which provided some game for the pot. On 8 February, again rising early, the expedition set out from Safieh, rounding the southern end of the Dead Sea and heading out of the Ghor for Kerak. Now north of the point where in 1864 Tristram had turned his back on the eastern side of the Dead Sea, their journey consisted mainly of geological observations and the photography of the ruins encountered. Nothing of the birds seen was recorded, except for odd references to the Greek Partridges which were shot for food (Tristram, 1873).

Before entering Kerak the expedition was met by horsemen from the town demanding seventy pounds for their safe passage which Tristram eventually negotiated for twenty-five Napoleons—a gold coin issued in the time of the Emperor Napoleon I, with a value of twenty francs. Once in the town, the local headman demanded £600 from Tristram for their safe passage and initially refused less. From the expedition's point of view this was potentially a more serious set-back than that experienced by Tristram some eight years previously. Unable or unwilling to pay the ransom, Tristram secretly sent an urgent note by messenger, through a local Greek schoolmaster, to the British Consul in Jerusalem appealing for help. Tristram then decided to take a firmer line, and, dressed in his full expedition regalia before the headman, demanded their release. In the meantime they had managed to acquire another guide for the next weeks at a cost of sixty pounds. In addition, the fact that Tristram had communicated with Jerusalem became known and the expedition decided to attempt to leave Kerak. This they unexpectedly achieved the following morning. Through driving wind and rain, described by Tristram as a "hurricane", they made their way to camp at Rabba. More geological observations were made but the weather prevented photography and no birds caught their attention until in a storehouse in Rabba they encountered "myriads" of

Rock Doves. In the meantime Tristram obviously forgot to communicate their departure from Kerak to the authorities in Jerusalem.

Leaving Kasr Rabba there were enough birds to distract Tristram from the local architecture: "Sandgrouse, dotterel and Plover, golden and Asiatic, were in plenty but rose wildly out of shot; and I saw some of the graceful Black-winged Stilt (Plate 52), allured by the smaller pools left here and there by the rains." The disagreeable experience with the natives as they journeyed north from Kerak obviously affected, to some extent, their ornithological observations and it was not until they were heading along the Ravine of Arnon that Tristram commented that pigeons and partridges abounded: "Partridges really swarmed; the lovely little Heye's Partridge, with its delicate plumage on the lower and warmer parts of the pass, and higher up the fine Greek Partridge, giving out its cheery 'chuckor—chuckor' from the top of every rock and boulder. An abundant supply for supper was easily secured." The Wadi Mojab or Arnon provided more ornithological interest.

> Whilst we stood on the edge looking down into that noble rift, the great birds of prey were sallying forth to forage. The Griffons circled and soared from their eyries lower down, till lost to sight in the sky; the buzzards lazily lapped their heavy wings as they crossed and re-crossed; but grandest ornithological sight of all, a pair of Lammergeiyers, the largest on the wing of our raptorial birds, kept sailing up and down, backwards and forwards, quartering the valley and keeping always close to the brow, the sinuosities of which they followed without a perceptible movement of their wings; only their long tail steering them in and out, as each time they passed us, easily within gunshot on a level with our eyes. They were perfectly fearless, as though they knew the sportsmen had only No. 7 (shot) in their barrels; and in the morning sunlight their brown tails and wings gleamed with a rich copper hue and their ruddy breasts shone brightly golden.

Tristram had not quite forgotten his birds.

North of Dhiban the party approached the town of Um Rosas which they made their headquarters for a week. Outside the town they encountered two great cisterns, one with a vaulted roof and still containing water. Here they found Rock Doves, Kestrels, Ravens, Jackdaws and owls (probably Little Owls). In the town Tristram commented: "Man has given place to Partridge of which the numbers had not perceptibly diminished by the end of the week though they had supplied our large party with two plentiful meals daily. This fine bird would never be found in the plains. Strictly a rock bird over the steep hillsides and cliffs of Syria everywhere, on the plains, wet or dry, it is never found."

There were many ruins on the surrounding plains and remnants of channels, dams and sluices which contained large numbers of white snails and red caterpillars, like those of the Emperor Moth. "Here, myriads of larks—the skylark (Plate 53), crested lark, short-toed, calandra and others in combined flocks (were) fattening themselves upon them. Here and there were a flock of rock doves fluttering from a cistern; a covey of partridges from a ruin; a pair of Egyptian Vultures fattening on the offal from a recent camp."

Sunday 17 February dawned with great embarrassment for Tristram. His letter to Jerusalem, asking for aid, had resulted in a message being sent with a letter from Mr Selami of the English Consulate. In short it stated that he was on the way to rescue them, together with the Pasha of Nablus, and troops, both horse and foot, with two brass canons. He was carrying £600 for the ransom and assumed that no force would be used until they were safely out of the way. Tristram who had forgotten to cancel his request for aid took charge and decided to return to Kerak to apologise. Without further ado, apart from a short morning service—it was Sunday—he set out back to Kerak over the plateau of Moab, where they observed cliffs with many open caves containing nests of Griffon Vultures and Lanner Falcons. Trips of dotterel were disturbed by the travellers and vast packs of sandgrouse "rose wild and fast as pigeons".

After ten hours in the saddle they found the camp of the Pasha and Mr Selami and were subjected, not surprisingly, to two hours of questioning. As a result, the Pasha countermanded, provisionally, the order for 500 armed men who were to follow him from Nablus as a reserve. The guard that he had brought with him to rescue Tristram and his friends were camped around, and consisted of "170 infantry, 120 cavalry, two field pieces [the brass canons] and 150 mounted irregulars"—enough to start a small war! In the end Tristram overcame his embarrassment and "departed the camp to a salute from the howitzers". He felt even better when subsequently he discovered that the Turkish government "had thrown a garrison into Kerak"!

The return to their own camp was uneventful and no mention was made of any birds encountered. The following morning a masked Arab rode into the camp and informed Tristram and Zadam (now returned) that his tribe were the rightful owners of the surrounding land and that "his gunners would come down in the night and shoot us all". Yet even more excitement if that were possible.

Supplies were running low as they left Um Rosas but pigeons and partridges were plentiful and easily obtained. The ruins of Ziza occupied the expedition for much of the following few days and then on the next Sunday Tristram records that "The Sakk'r Falcon sat calmly on his favourite perch and allowed us carefully to reconnoitre him . . . while the eagle owl, sandgrouse and partridges showed a similar contempt for unarmed Europeans"—unarmed because it was Sunday.

Moving on from Ziza to Mashita (Um Shita) the only ornithological recordings were made in the ruined palace where, on an octagonal tower, there were interesting carvings of peacocks, partridges, parrots and other birds. It was decided to spend a second day at the ruins around Mashita and at the ruins of Rustal game was abundant: "It was curious to see Mallard and Pintail feeding amongst the stunted scrub in most unlikely ground for duck. A fine imperial Eagle sat on the carcass of a kid until I was within ten yards of him, showing his white shoulder in fine contrast with his dark plumage . . . But I had only small shot!"

"In the evening I found that Trotter, riding in another direction, had shot on the plain a duck which proved to be a hybrid between mallard and pintail, the facsimile of the so-called Bimaculated Duck of Yarrell's British Birds." (Yarrell, 1871) The bird included in Yarrell's description is said by him to be a hybrid of the Mallard and Common Teal (Plate 54), but the plates featured in both Meyer (1835) and Bewick (1847), and labelled Bimaculated Duck, are clearly of the Baikal Teal *Anas formosa*.

Abroad the following day Tristram estimated over one hundred Griffon Vultures on the carcass of a camel and when disturbed they "darkened the air". That day was the first sign of spring with Hoopoe (Plate 55) sighted and swallows entering one of the tents. Reaching Sufa the party encountered a falconer with two Saker Falcons "well known and distinct from the peregrine and the lanner, which latter our falconer ensured us was a very sluggish falcon and worthless for gazelle." He was not tempted to sell his birds at any price and treated Tristram's offer of "£10 apiece" with scorn. Tristram claimed that his whole collection had cost less than £100 (Tristram, L. H-H., 1898), but this was probably for Mrs Tristram's benefit.

The following day, 29 February, Tristram's party received a great honour, a visit from Fendi y Faiz, Zadam's father and the great sheik of the Beni Sakk'r. Tristram purchased two sheep for the occasion and commented: "A visit from a king is not an everyday occurrence and it required all our stock of dignity, coffee (Mocca) and tobacco to receive Fendi and his three sons all at once." As though this were not sufficient excitement for one day, as soon as it was dark they heard "not only of departures but of war". Fendi el Fez was off to conduct and defend a party of pilgrims on their way to Mecca and at the same time Zatun, Zadam's brother, was off eastwards to avenge the robbery of camels by the 'Amzeh.

After they moved further north, the transition from the highlands to the mountains was very sudden and the climate and vegetation changed too. The cliffs afforded many attractions for naturalists and Tristram turned his attention once again to the birds. They found the nest of a Spotted Eagle with the bird sitting and out of reach, and "for the first time we heard the cuckoo's note resonating in all the glens." Overhead were Alpine Swifts, but again out of reach of Tristram's gun. As the gorge narrowed the newly encountered basalt lay horizontally on the limestone, and warm sulphurous springs (Solomon's Springs) issued from the ground. This turned the interest of the expedition to the local geology. Soon, however, Tristram encountered all the birds "which had been the novelties and prizes of my first expedition to the shores of the Dead Sea—the short-tailed (Fan-tailed) Whistling Raven

*C. affinis*, the bulbul, the bush babbler, the Moabite Sparrow (Plate 50) and many a rare warbler inhabits the thickets or scans the cliffs above them."

Based around the hot springs they made expeditions in several directions and saw many rare and interesting birds. "Several pairs of Tristram's Grackle, with their ringing, bell-like whistle showed themselves, but, when shot, fell hopelessly into the abyss; and a large flock of over a hundred of the rare Wedge-tailed (Fan-tailed) Raven *C. affinis* wheeled for half an hour over our heads as we descended into Callirrhoe."

From Callirrhoe the expedition followed the feeders of the river into the mountains and from there their objective was to descend westwards towards the Dead Sea, which they reached at Zara, three miles south of the mouth of the Callirrhoe River. Here, in what Tristram described as "eternal summer", they encountered Lesser Whitethroats, Blackcaps and Chiffchaffs as they made their way north following the river back to their camp. So strenuous was the journey that all that was recorded was its difficulties. Afterwards Tristram described it as "the most successful, original and enjoyable day we ever had"—so he must have seen some birds along the way. The party stayed two more days in the camp, hunting ibex, dealing with the collections and with photography. As it was, they had little time for birds. Tristram's Grackle, referred to here by Tristram as the Red-winged Grackle, was common on the cliffs and Moabite Sparrows, the little sunbird and a rookery of "square-tailed" (Fan-tailed) Ravens passed over their heads in the morning and evening making their way to and from the roost on the basalt cliffs. Tristram found this last species interesting to watch, its behaviour Jackdaw-like, in flight dipping vertically and performing somersaults in the air.

The expedition at this stage seemed to be intent on searching out the remains of old civilisations, particularly old ruins, to the exclusion of all else, and four days were devoted to the remains around Mediba. They were surprised by a visit of Jericho Arabs, amongst them Tristram's former henchman Jameel, who had collected specimens for him on previous visits. Jameel was provided with powder and shot, arsenical soap and carbonic acid (for preserving skins) and instructed to meet them again in five days' time. This he did and provided many specimens for them, but Tristram was disappointed by his skills in taxidermy. For some days afterwards it was ruins and yet more ruins, drawing ruins, photographing ruins and discussing ruins. Arriving in the neighbourhood of Madaba (Mastaba) rides in the area apparently yielded nothing of ornithological interest worth recording.

North of Madaba Tristram encountered what would have been a good camping ground, had they previously known of it, and he explored the area for an hour. "Colonies of (Lesser) Kestrels were debating overhead, the fine Alpine Swifts had already returned and dashed, shrieking from their inaccessible chinks, and a rookery of Square-tailed (Fan-tailed) Ravens were amusing themselves by the futile pursuit of a Lanner Falcon." Further on, in a cleft in the rocks, they passed the Wadi Sakk'r, clad with palm trees. "Appropriately it contained a pair of Sakk'r Falcons swooping round and round overhead as we approached too near their nest on the cliffs . . . [but] the nest was inaccessible." A second, equally inaccessible nest of another pair of the falcons was present further along the ravine. Tristram left the main party and went off by himself, as he was often inclined to do, and late in the afternoon the main party caught up with him.

> Birds of all kinds, rare and common, tantalised us as we rode, for we had no time to shoot. The summer migrants were travelling northward; flocks of Great-spotted Cuckoos, scores at a time, rose with their steady flight, waving their long tails as though they could scarcely wield them; all our English songsters were mingled with bulbuls, sunbirds and the rare denizens of Jericho. Now and then a desert hare, and every few steps a quail, started beneath our feet, while partridges, Greek and Heye's, called defiantly from the tops of boulders as we passed, as though they knew themselves safe.

Back in the camp they found the tent turned into a museum by the trophies obtained by Jameel. "Birds, beasts, eggs, reptiles, beetles and shells lay strewn in profusion. Among them . . . grackles, hoopoes, spotted cuckoos, hares, hawks, desert partridges by the dozen . . . all hinted by their perfume that they must be either stowed away or thrown away, without delay."

Their last night in Moab "was not without ludicrous adventure". Tristram alone, the rest of the party having taken to their beds, was at work with his carbonic acid and arsenical soap, when the wind increased fiercely and with a sudden gust the second tent was blown away completely. Fortunately the photographers and botanists had stowed their equipment but "the plain was strewn with the debris of the wreck".

In the morning, after clearing the wreckage, the expedition decided to take the new ferry across the Jordan in the company of "an arsenal of a dozen flintlocks [which] indicated that the thickets of the Jordan are still the haunts of outlaws and robbers who have fled from justice." The following morning they spent in bird-collecting in the thickets and examining the excavations of Old Jericho which were financed by the Palestine Exploration Fund (PEF). Setting out from the remnants of the ancient city the expedition passed through Bethany and into Jerusalem, but not without a fight between the muleteers as to which had proved itself the best mule. "The place of honour was won (surprisingly!) by a Christian mule from Lebanon." Great was the indignation of the devout Muslims, but forgotten by the evening, when with Zadam as their guest, whilst he complacently quaffed his champagne (presumably non-alcoholic), they ended their wanderings in the Land of Moab.

— 11 —

# Palestine Revisited: Egypt to Armenia (1881)

EGYPTIAN VULTURE. (*Neophron percnopterus.*)

In 1881, twenty three years after his first visit to Boulac, originally a little fishing village less than two miles from the centre of Cairo, Tristram found himself there once again. After such a period of time things have often changed and he was disappointed to find few birds present. Gone were the Avocets, Spoonbills and waders of his previous visit and in a day's journeying by steamboat he saw few birds along the river. His first visit was in March 1858 and at this time migrating birds would have begun their northward journey so he was probably expecting too much of a January visit (Tristram, 1882).

Tristram spent a few days in Egypt immediately prior to his fourth visit to Palestine and as his studies of birds of the Bible had led him to speculate on time spans affecting the bird life, he found some differences in those birds depicted in the temples and tombs of the past, compared with his daily observations. Of particular interest to him was the depiction of two "Red-breasted Geese of Siberia" in a line of six geese (Plate 56) drawn life-size on a fresco from "a tomb of the Hyksos period" then in the Boulac museum. Tristram identified the other four birds in the fresco as White-fronted Geese, and he speculated that this piece of ancient art perhaps indicated that Red-breasted Geese bred further west at the time of the painting, which is now referred to as the mastaba of Atet, Meidoum, Dynasty IV. Tristram's examination of the fresco was perhaps cursory and in poor light, since he very unusually failed to note that, of the other four birds, only two were White-fronted Geese, the two end birds being Bean Geese (Newton, 1896; Barrett-Hamilton, 1897). It is surprising that Tristram made this mistake as he was the first to report Bean Geese in Palestine (Shirihai, 1996), having seen them available in the markets. His own collection Catalogue (Tristram, 1889) indicates that he had shot the species in Malta in 1858 and G. H. Gurney had shot a single Bean Goose at Greatham whilst staying with Tristram. Had he identified the Bean Geese on the fresco, Tristram would probably have shown as much interest in them as in the Red-breasted Geese as, according to Shirihai (1996), the Bean Goose is the rarer of the two species in Palestine, and in Egypt there are only two records for the nineteenth century and one for the Red-breasted Goose (Goodman and Meininger, 1989).

Tristram's party, the membership of which he does not make clear in his account of the expedition, left Cairo for Ismalia and Port Said where birds were abundant but he secured only a single Avocet, which is not recorded in his Catalogue. Once in Palestine the party moved quickly south to Beersheba, seeing many of the species of birds which Tristram had seen on previous visits. The cranes were moving north during February and he spent much time watching them but he was never able to get close enough to add one to his collection. He concluded that the species must be an early migrant as he did not see a single bird after the end of February.

It was not until the beginning of April that the northerly migration began in earnest, though Swallows were passing north throughout February and the first Swifts were seen at Madaba on 27 February. No details of the journey between Beersheba and Madaba, near the north-eastern end of the Dead Sea are given, though it seems probable that the party proceeded north along the shore of the Dead Sea as no mention is made of Jerusalem. The movement of Alpine Swifts took place throughout the next day and no more were seen until they were found nesting further north. The seventh of March saw the main movement of Common Swifts "overspreading the whole district and remaining for nidification".

On 19 March Cretzschmar's Bunting was found spread in pairs over the hillside but it was not until 5 April that the much more common Ortolan Bunting arrived. This species did not pair off until a fortnight later when Cretzschmar's Buntings were sitting on eggs. The Black-headed Buntings arrived on 7 May and were the commonest breeders of the three species of bunting. Some White Storks arrived on 10 April, the passage was complete by 22 April and none were seen after this date. On 16 April the House Martins passed north in their thousands and the same day a vast cloud of Swifts, quite apart from those which had overspread the land, dashed in the same direction up the valleys.

Lesser Grey Shrike, which Tristram had managed to obtain only once in 1856 and had not seen at all in 1863–4 and 1872, did not arrive until the middle of May but remained very common after this date, whilst the Masked (Nubian) Shrike made its appearance on 11 April. At the beginning of April, on the low ground near the source of the Jordan, Tristram found several Snow Finches, enabling him to confirm his belief that it was the same species as the western Alpine birds. In the 1863–4 expedition to Palestine the party did not record or collect a single Collared Flycatcher but in 1881 they were everywhere, whereas only one Pied Flycatcher was seen. "Not a Flycatcher did we see till, on the morning of 16th April, crossing from the stupendous gorge of the Leontes to the head of the Jordan valley, and then up the woodland to the Banias, a bright little black and white bird, conspicuous amongst the foliage and suggesting the male, started from about every other tree and often from the scanty scrub."

Shirihai (1896) draws attention to the great variation in the spring numbers and arrival dates between years: "very great annual variation in number and peak dates; in some springs only a few appear locally, but in others tens of hundreds occur in various areas." The same author comments on the similarity of the variations in numbers between Collared and Pied Flycatchers. However, the latter arrives in two peaks, 13–25 April, with males arriving first, females towards the end, and 2–7 May, where females predominate. Collared Flycatchers also arrive in two waves but these are not so distinct as those in the Pied Flycatcher. Clearly 1881 was a good year for Collared Flycatchers and 1864 for Pied Flycatchers. Tristram also noted that the birds in 1881 did not appear to be paired. Of the Collared Flycatchers in 1881 Tristram commented: "The whole land was covered with them. And this continued as we travelled eastward and southward up to 29 April. Everywhere a male bird was seen; but they certainly were not paired."

Another migratory species which particularly interested Tristram was the Great-spotted Cuckoo. Unlike the European Cuckoo, this species migrates in flocks. Tristram encountered such a large flock on 22 April and he had previously come upon a similar roosting flock in the Jordan valley on his 1872 expedition: "They travelled very leisurely, and whilst scattered along the whole length of the valley which they were crossing, kept up a ceaseless conversation."

Tristram (1882) refers to the fact that there had been critics of his separation of his Mountain Short-toed Lark *C. hermonensis* and the Large Short-toed Lark *C. brachydactyla*. Nowadays the two are regarded as conspecific though the former is recognised as a distinct sub-species—*C. b. hermonensis*. Tristram was a very good field biologist and in this he had an advantage over many of the contemporary taxonomists. A bird in the field has very different characteristics from a skin on the bench and Tristram (1866b) appreciated this. *Calandrella hermonensis* appeared to be confined to the higher mountain zone of the north, and to be a permanent resident. Breeding three weeks earlier than the *Calandrella brachydactyla* he was familiar with, it appeared larger, had a longer and more slender bill, a bright rufous colouration and a distinctness of the blackish collar, with a much smaller extent of white on the outer tail feathers. Tristram (1882) further pointed out of *C. brachydactyla* that its haunts are in the plains whilst *C. hermonensis* was found only in the rocky heights; its flight was quite different and Tristram thought it impossible for anyone who had been once introduced to both species in their own home to confuse them. *C. hermonensis* he described as perching on the edge of a rock or on top of a small boulder and uttering a repeated, rather plaintive, but very clear note, utterly unlike that of the Short-toed Lark, and something like that of an exceptionally musical Yellow Hammer. Also he recorded it as being solitary, and not a gregarious bird, rising up into the air much like a Skylark but not attempting to produce the lark's sustained song for any length of time. Tristram recognised differences which many taxonomists did not because, firstly, they had not had the opportunity to see many of the species whose skins they handled in a field situation and, secondly, this was the age of the morphospecies—the species defined by the morphological differences which were recognised only in the type specimen(s). Thus Sharpe (1890) concluded: "*Calandrella hermonensis* of Tristram cannot be upheld as a species as such as it decidedly grades into *C. brachydactyla* as regards every character in which it is supposed to differ from that species." This statement, whilst it is correct for morphological characteristics, takes no account of the differences found by Tristram in the field. In the long run this may prove to be a correct decision and the differences which exist are recognised in the sub-specific designation of *Calandrella brachydactyla hermonensis*. Eight sub-species of Larger Short-toed Lark are at present recognised (del Hoyo et al., 2009) but future work and the advances of molecular techniques may well result in changes. It may not have been recognised at the time, but Tristram was helping to move forward the fundamental unit of taxonomy from the individual to the population, so further bolstering Darwinian theory. In this Tristram was before his time in anticipating the central position of the population.

Whilst only seven of the "new species" which Tristram described are still recognised, another forty remain as distinct sub-species or races and possibly some of these may regain specific status in the future. In the list of Types in the Liverpool Museum (Wagstaffe, 1978), the Greater Short-toed Lark is listed incorrectly as *Calandrella cinerea hermonensis*. *C. cinerea* is the closely related Red-capped Lark of southern Africa, and Tristram's "Mountain Short-toed Lark" is *Calandrella brachydactyla*.

During the whole of the 1881 expedition Tristram found only four species of birds which he had not encountered before in Palestine: Pied Wheatear, Goldcrest, Great Snipe and Bearded Tit. Tristram was always on the look-out for changes in comparison with previous visits and in 1881 he found the francolin to be much commoner and widespread. Extinct in Europe, "in Cyprus the annexation to England sealed its doom there", in Syria it was positively thriving and had spread over the whole of the lowlands north of the Lebanon.

Tristram was also intrigued by the sharp line of demarcation between species and that a range of mountains such as the Lebanon, rarely exceeding 10,000 feet, should prove such a barrier. Whereas in Palestine the "Rufous Warbler *Sylvia galactodes*" was very common, he found that after crossing the mountains it was replaced by the "Grey-backed Warbler *Sylvia familiaris*". Nowadays they are regarded as conspecific, but as in the case of the Short-toed Lark his species are given sub-specific status. Now the birds are known as the Rufous Scrub-Robin *Erythropygia galactotes, E. g. galactotes* being Tristram's Rufous Warbler and *E. g. syriaca* being the so-called Grey-headed Warbler.

On the Syrian plains Tristram again had difficulty with the nuthatches. He recognised a smaller form which he refers to as *Sitta neumeyeri* and a larger form which he refers to as *Sitta syriaca*, which he points out is not found in Syria. *S. neumeyeri* he describes as being found in all the mountainous regions of Galilee and Syria whilst his *S. syriaca* was to be found only in Armenia. However, they differed morphologically only in size and Tristram writes: "I admit them to be distinct races, but grudge specific distinction to mere size, especially when, as here, I could detect no difference in the voice."

Present day taxonomy (del Hoyo et al., 2009) recognises six races of *S. neumeyeri*, one of which, *S. n. syriaca*, is that occurring in Israel and two others, which Tristram may have encountered in Armenia, being *S. n. tschiescherini* and *S. n. plumbea*. Either of these would meet Tristram's criterion of similar voice but neither of them that of size difference. This and the fact that the two forms of differing size were breeding in the same geographical area, i.e., were sympatric, suggest specific difference, and by the time that he came to write his magnum opus, *The Natural History of Palestine* in 1884, Tristram had come to the same conclusion, though he still refers to the larger form as *S. syriaca* in his comments on *S. neumeyeri*. This species is common all through Lebanon, whilst north of the mountain "I found both this and the true *S. syriaca*, the large species identical with that of Central Asia, inhabiting the same localities." Here Tristram is referring to the nuthatch now named *Sitta tephronota* described by Sharpe (1872). This is now accepted as the Great Rock Nuthatch. Dresser (1871–90) did not accept this and wrongly considered Tristram's *S. syriaca* and *S. neumeyeri* as synonymous. Again this shows what a good field observer Tristram was. He certainly appreciated the differences between the large and small nuthatches and eventually, after careful consideration, came to the correct conclusion.

Tristram was clearly interested in boundaries between different populations. He commented particularly on the boundaries of the Magpie population just north of Aleppo where he considered the area to the south to be similar habitat. However, his thoughts turned elsewhere when he came upon "the most interesting ornithological sight I saw in Syria"—the Rose-coloured Pastors on migration. At the time he was standing on the battlements of the citadel of Larissa, some sixty feet above the gorge. A dozen pairs of Lesser Kestrels were circling the area and he was watching a flock of European Bee-eaters digging out their nests in the adjacent river bank. Rock Swallows and a pair of Wall Creepers were also in his area of observation and Rollers were noisily chasing away a flock of Jackdaws.

With a rush of wings and a darkening of the sky a massive cloud of birds dashed through the river gorge, out onto the plain. Looking down on the birds the members of the party could clearly see their rosy backs—they had come upon the migration of the Rose-coloured Pastor (Plate 57). A second cloud appeared and dropped on an islet in the river; the islet suddenly turned from green to black. Tristram fired once at random and an hour later picked up five birds in full breeding plumage and later a second single shot afforded ten more skins.

As they followed a north-easterly course across the Syrian plain for two whole days, flock after flock crossed their path flying westwards, probably, as Tristram pointed out, in search of locusts, all the time emitting a continuous loud chatter. As they crossed a patch of vegetation recently devastated by locusts, it was seen that no adult insects were present, but recently hatched young locusts were everywhere. Once past the devastated area they turned to see a flock of Pastors descend upon the area and after watching them feed for a short time the party returned to the area to find it cleared of young locusts; the Pastors had consumed all of them. After three days of continuous passage Tristram recorded that he did not see another Pastor—they all had gone west. It was then near the end of May and in the flocks Tristram had not detected a single young bird, and it was clear that they had not bred!

From Anitab Tristram struck down the Euphrates to Birejik where he was hoping to find Bald Ibis. This species the party saw for the first time whilst waiting for the ferryboat that evening. They had been watching Persian and European Bee-eaters on the telegraph wires and Redshank (Plate 58), Green Sandpiper and Kentish Plover (Plate 59) by the riverside, when long lines of black birds, like ravens in the distance, flew very low over their heads to the far side of the river; "and we recognised them at once as the great Bald Ibis (*Comatibis comata*), whose acquaintance I had made only once in my life, in the Sahara." The birds were in easy shooting distance but Tristram decided to contain his impatience to add to his skin collection. He had been told that the birds were sacred and it would be a crime to kill one as the local inhabitants believed that the ibises carried the souls of departed saints. Not to be dissuaded by the local folklore, a day or two later Tristram hired a local guide who apparently was adequately rewarded to ignore the statuary by-laws and he was led to the nesting area at dusk. As the birds quietly sailed home at sunset, he quickly picked off six birds, three of which are recorded in the Catalogue of his collection. Tristram subsequently wrote: "I have no desire ever again to skin six ibises after a hard day's work, by lamplight, in a temperature of nearly one hundred degrees."

Tristram recorded that he frequently flushed Black Storks whilst walking along the banks of the Euphrates, and at Carchemish he was able to indulge his interest in archaeology without neglecting his birds. Here, in sixteen shafts dug by workers from the British Museum into the great Hittite mound, both European and Persian Bee-eaters made their nest tunnels side by side and the shafts contained colonies of both species. Whilst the European Bee-eater hunted high in mid-air and over the river, "its congener at once took himself less adventurously to the plain and ruins beneath, and there skimmed close to the surface, perching continually on the stones which strew the site of the ancient metropolis."

Crossing the Euphrates at Samosait, the expedition entered southern Armenia and encountered a very different flora and fauna than that which they had left behind. "Instead of the Isabel Wheatear (*Saxicola isabellina*), which with the Calandra and Short-toed Larks was almost the only winged denizen of the plains, where it is in amazing numbers, every turn every clump of trees now introduced to us some old or new feathered friend." Between Beshni and Nadjar, at a height of some 7,000 feet, (2,134 m), Tristram commented that the variety of birdlife was bewildering. Here he obtained a specimen of the Cinereous Bunting, a species which he had not previously obtained, feeding in small groups and concealing itself in the scrub. Also present were the two species of Rock Nuthatch, White-throated Robin, Sombre Tit, Middle-spotted Woodpecker and many Magpies and Jays. Here the oologist came out in Tristram: "I cannot conceive a field better likely than this to reward an oologist early in the season. Every Warbler, Redstart and Bunting of Eastern Europe seemed to abound."

The account of this expedition is completed by Tristram referring to his experiences on the Lake of Antioch, in northern Syria. This was a shallow lake, some miles in extent and swarming with eels. On its northern side were countless islets on which Darters (Snake Birds, Plate 60), Little Cormorant and Common Terns were breeding. That the last of these was present and breeding surprised Tristram in the absence of White-winged Black Terns and Whiskered Terns which he asserted would have been present in similar circumstances in Algeria. The nests of the Darters particularly interested him in that each was little more than a down-trodden tuft of grass, flattened by use, each nest being very close to the next. The eggs of the Darter were laid very late in the season and did not hatch until June. As soon as the young were able to fly, Tristram observed that the whole colony rapidly disappeared and did not return to the nesting site until the end of the following April, a short time later than the Little Cormorants.

Tristram elaborated no further on the ornithological aspects of this expedition, the highlight of which, he wrote, was his observations on the Bald Ibis. He clearly liked Palestine very much for several different reasons and he was destined to make two more visits, in 1894 and 1897. Whilst it is unlikely that Tristram neglected the birds on these later visits to Palestine they are not documented in the way of his previous visits. His leg was broken by a horse kick during this last visit and although he seemed to recover quickly, it must have restricted his activities. The Catalogue of his collections published in 1889 obviously cannot contain any records of those visits and neither can his *Fauna and Flora of Palestine* (1884) for the same reason. Probably he considered his ornithological interests in Palestine complete after these publications and his interest had moved to pastures new in that he was much involved in the ornithology of oceanic islands and also that of Japan. Somewhere he also had to fit in his canonical duties—one wonders how.

Figure 7: Map of the Dead Sea area. From H. B. Tristram,
*The Land of Israel*, 3rd rev. edn., 1876.

Plate 56: The Mastaba of Meidoum (IV Dynasty), remarked on by Tristram, actually features three species of geese, Red-breasted, White-fronted and, the two end birds, Bean Geese, suggesting that they were commoner then than now.

Plate 57: Rose-coloured Starling *Sturnus roseus*. A dramatic migration of vast flocks of these birds, consisting entirely of adult birds, was seen.

Plate 58: Common Redshank *Tringa totanus*. Seen with other waders on the banks of the Euphrates, near Birejik.

Plate 59: Kentish Plover *Charadrius alexandrinus*. Many were observed scurrying across mud banks on the Euphrates.

**Plate 60:** Darter *Anhinga rufa* was found nesting commonly on islets in the Lake of Antioch in northern Syria. From Tristram's *Fauna and Flora of Palestine*.

Fig. 66.—Dodo. *Didus ineptus.* (After Savery's Vienna picture.)

**Plate 61: Dodo** *Raphus cucullatus.* The *Didus ineptus* of Linnaeus was of particular interest to Tristram because of its flightlessness and the fact that it was extinct. His collections contained several remains of the bird obtained through his friend Edward Newton, brother of Alfred. From Newton's *Dictionary of Birds*.

**Plate 62: Common Kestrel** *Falco tinnunculus.* A species with a range from Spain to Japan across areas occupied by other species of Kestrel. This image of a hovering bird was taken by Phil Collins.

Plate 63: Vegetarian Finch *Platyspiza crassirostris*. One of the first species of Darwin's Finches to be described by Gould.

Plate 64: Woodpecker Finch *Camarhynchus pallidus*. One of the few species of birds to use a tool.

**Plate 65:** South Island Moa *Dinornis robusta*, originally found buried in a sand drift with broken eggs, this specimen retained skin and feathers according to Tristram (1877). It was unfortunately "restored". Image by W. G. Hale reproduced by courtesy of the York Museums Trust, http://yorkmuseumstrust.org.uk.

Plate 66: Takahe *Porphyrio mantelli*. Thought to be extinct this flightless rail was found in Fjordland, South Island of New Zealand in 1948.

## — 12 —

# Tristram's Islands

CORMORANT. (*Phalacrocorax carbo*)

For anyone first visiting the Farne Islands it is an experience never to be forgotten. For Henry Tristram the schoolboy the Farnes were a magical place and remained so for the rest of his life. His time in the Bermudas was really another adventure for him, and then, from Newton and Wolley, the talk of the Great Auk Islands held a fascination for him. Particularly they talked of Eldey, off the south-east coast of Iceland, but he was never to visit it. Similarly, he heard about many other islands from the missionaries he sent out to collect birds and eggs for him; and from his long discussions and arguments with Newton he came to appreciate that the fauna and flora of islands close to land masses were very similar, whilst those of oceanic islands distant from the nearest continental masses differed significantly.

Throughout his life he thought more and more about this and went to great trouble to acquire specimens for his collections from islands across the world. However, he visited few, only Sicily and the Canaries, towards the end of his travels. Whilst most of his many publications on birds appeared in recognised scientific journals, such as *Ibis* and the Proceedings of the Zoological Society, Tristram published in a variety of very unusual journals, several of which no longer exist. These include *Good Words, The Contemporary Review, Christian Advocate, Clergyman's Magazine, Leisure Hour, Sunday Magazine, Sunday at Home, Good Words for the Young, Church Sunday School Magazine, Evening Hours, Mission Life*, and *Churchman*. He selected *Good Words* in which to publish his thoughts on islands under the eccentric title of 'The Story of the Isles of the Sea, told by the Fowls of the Air' (Tristram, 1877). He chose to introduce his account unusually by selecting a botanical illustration—the *coco-de-mer*, a form of very large double coconut first encountered by sailors in mid ocean between Africa and India, long before there were proper charts. Because of its rarity it brought enormous prices in India, as it was thought to have extraordinary medicinal powers. It got its name because no one knew from where it originated. Then, in 1742, a French ship from Mauritius visited a group of islands originally discovered by the Portuguese and subsequently known as the Seychelles. Here they found the *coco-de-mer* growing on some of the uninhabited islands, but later in Tristram's time it was confined to one of the smaller rocky outcrops of the group, the Isle de Praslin, and to another insignificant rock. Having found its origin, the *coco-de-mer* lost its curiosity and its so-called medicinal powers and became worthless. However, it served as an example for Tristram of an organism whose distribution was limited to a very small group of islands and was completely unknown on other islands and the adjacent continental land mass. This was in marked contrast with other species of palm which have a very extensive distribution. Tristram realised that this sort of distribution was not limited to plants but that other organisms were similarly distributed and even some species of birds were limited to particular islands.

In his *The Origin of Species* (1859) Darwin summarised the then-known differences shown by the distribution and variation of different groups of organisms on islands, suggesting that the reptiles on the Galapagos Islands and large flightless birds in New Zealand occupy "or recently took" the place of the absent mammals. Newton, like most other Victorian ornithologists, took an obsessional interest in the Great Auk (Wollaston, 1921), and particularly the fact that it was flightless. This led to his becoming interested in other flightless birds, for example the Dodo, many of which occurred on islands, and this led Newton's biographer Wollaston to comment that island faunas, particularly of oceanic islands, "had a peculiar attraction for him"! Of course Newton and Tristram exchanged views on just about everything but both were particularly interested in the flora and fauna of islands generally, and when Newton's brother Edward—another keen ornithologist—was appointed in 1859 to be Assistant Colonial Secretary to Mauritius, this gave both access to specimens taken from that island and from others of the Mascarene group. Somewhat later Edward Newton's ornithological work in the Mascarene Island attracted the attention of the British Association and they appointed Tristram, together with Alfred Newton and Philip Lutley Sclater, to "a Committee for the purpose of assisting Mr E. Newton in his researches for the extinct Didine (related to the Dodo) birds of the Mascarene Islands." (Wollaston, 1921)

## The Indian Ocean: the Mascarene Islands

The Mascarene Islands are true oceanic islands, not being situated on the continental shelf of the adjacent land mass and being of volcanic origin, the latter characteristic being known to Tristram. What was not known at the time of his writing was that Mauritius, Réunion and Rodriguez are connected by the undersea Mascarene Plateau which at different times and different sea levels may well have connected the three islands with the Seychelles group which lie some 1,100 miles (1,770 km) to the north. However, in his writings Tristram associated the Seychelles with the other Mascarene islands and speculated that organisms inhabiting such islands would be those that drifted on currents or, for aerial organisms, were those blown off course. He had experience that this was the case in the Bermudas, they too being volcanic islands separated from the nearest continental shelf. However, in the latter case the islands possess no endemic birds whereas they are commonplace in the Mascarenes. As in the Bermudas, there have been no flightless forms of birds recorded in the Seychelles though several species, both living and extinct, have been recorded in the distant Mascarenes. Tristram recorded some thirteen endemic birds on the Seychelles, eight on Mauritius, two on Réunion and three on Rodriguez. This compares with fourteen now recognised on the Seychelles, though two of Tristram's species have been reduced to sub-specific status, two are now extinct and three new endemics have been recognised. On Mauritius eight endemics are recognised, one of Tristram's being reduced to sub-specific status and one extinct; on Réunion seven endemics are now recognised and one of Tristram's is extinct; and on Rodriguez two endemics identified by Tristram remain, the third being extinct (Table 7). Elsewhere in the western Indian Ocean Aldabra boasts five endemics which were unknown to Tristram.

Species related to the island endemics exist on the adjacent land masses but these are different species and it clearly occurred to both Tristram and Newton that the differences were in some way related to the island existence. In his initial consideration of this, Tristram refers to the lack of any indigenous mammals on the islands and correctly interprets this as an indication that the isolation of the islands themselves over a long period of time was the reason for the continuing existence of such flightless forms as the Dodo (Plate 61) and Solitaire. For these he postulates "some generalised parent form which, by lapse of ages and differing conditions, had become stereotyped into the species of each island." This is a considerable step beyond natural selection within a species which, by the time of Tristram's publication of his thoughts in 'The Story of the Isles of the Sea', was still the limit of acceptance of Darwinism by some scientists. Before considering this matter further it is necessary to examine Tristram's objectives in writing this paper.

Firstly, it is unclear to whom the paper is addressed. Certainly it is to an audience wider than his more scientific publications, one more used to serialised novels by such well known authors as Trollope and Kingsley but one which would not be in any way critical of the scientific content of his writing. Secondly, Tristram wrote in a manner which lacks the clarity of his books and more scientific papers, and leaves much of the interpretation of his meaning to the reader. His introduction to the Mascarene Islands is particularly obscure: "We are arrested by the phenomenon that, geologically modern as might be islands of volcanic origin, there is here neither plant, bird, land-shell or insect, common to the nearest continent of Africa, nor to the most distant Indian regions on the other side. Every aboriginal is an original peculiar and unmistakeable; in the case of the birds the greater portions of the denizens being incapable of flight, could never have crossed from other lands." Here Tristram is clearly writing about land birds but this would not be interpreted as such by many non-scientists; in the case of the birds on what basis does he draw the conclusion that most were incapable of flight? There is no evidence that flightless forms ever existed in the Seychelles and all thirteen of the extant endemic forms are well capable of flight. The very fact that closely similar species occur on adjacent land masses would suggest that this was their origin. It seems, however, that Tristram was much taken by the fact that at least two species of flightless birds were once present on the more southerly islands of the Mascarenes, the Dodo *Raphus cucullatus* on Mauritius and the Solitaire *Pezophaps solitaria* on Rodriguez. The past existence of these two species is generally accepted on the basis of the evidence available, but verbal, written and illustrative accounts of other flightless species such as the White Dodo of Réunion and Leguat's Giant (probably a Flamingo) of Mauritius, quoted by Tristram (1877), are to be discounted as there is no good evidence that they ever lived.

Tristram (1877) records what he considered to be evidence for the existence of a Dodo-like flightless bird on the island of Réunion. When in 1642 the French took possession of the island, it was claimed that there too was a large wingless bird. The White Dodo of Bourbon (formerly the name of Réunion) was described by Du Bois in 1674 and a painting by Pierre Witthoos still exists. An early illustration by Bontekoe from about 1646 is recorded by Tristram as is a visit recorded by Tatton in 1613, who spoke of "a great fowl of the largeness of a Turkie and so short-winged that it cannot fly, being white". Bontekoe named it "Dodeersen". Tristram also records Carre, a French voyager, speaking of the beauty of its white and lemon plumage. As evidence that this might be an accurate description is the illustration made by Witthoos (died 1693) which is reproduced in the Transactions of the Zoological Society VI (Plate 62). Tristram describes the accuracy of the illustrations of other birds on the same painting by Witthoos, "an admirably exact drawing of a Red-breasted Goose and a Goosander". There are, however, no traces of bones on Réunion to support the validity of the painting, largely because, suggests Tristram, the French having eaten the birds fed the bones to the dogs, being "less considerate of the curiosity of posterity than the founders of the kitchen-middens of Denmark".

With the exception of the Dodo and Solitaire it appears that all of the flightless birds described in the old literature were almost certainly rails. The Mauritius Red Hen does have skeletal remains in support of its existence, as does a second rail with a decurved bill but since both are described as the size of a chicken, neither can be Leguat's Giant. More recently extinct was the large flightless Coot *Fulica newtoni*, named after Edward Newton (Alfred's brother), who collected remains from cave deposits in Mauritius. However, none of the reconstructions from these remains suggests a bird as large as the Dodo or Solitaire, though in each case they are large in comparison with their flying relatives, supporting Tristram's contention that flightless island forms tend to become larger. Only one flightless bird now exists on the islands of the Indian Ocean, the White-throated Rail of Aldabra; and other flightless birds, now extinct, were probably all rails, all birds much smaller that the Dodo and Solitaire. Wherever flightless birds occur they do so in the absence of major predators, particularly mammals. Tristram draws attention to the fact that the powers of flight, as judged by the size of their wings, are generally less in island birds than in their close relations on the mainland and this is demonstrated by the land birds of the Seychelles. Where this occurs there tends to be an increase in size (mass) of the birds concerned so that their movements are slower, which in turn results in their easier capture and demise should predators be introduced to their habitat. This has occurred in many islands throughout the world, resulting in the ultimate extinction of many flightless forms. In some cases, as Tristram points out, associated with the increase in overall size of some of the birds with reduced wing size, this is also associated with increase in bill size. Such is the case in the Turtle Dove *Streptopelia picturata rostrata*, a form recognised as a distinct species by Tristram in the Seychelles but now afforded only sub-specific status of the Madagascar Turtle Dove. Whilst the greater part of three continents is occupied by the European Turtle Dove *Streptopelia turtur*, the Indian Ocean has its own species which is at present divided into five sub-species. Similarly the Kestrel *Falco tinnunculus* (Plate 62) ranges from southern Spain to Japan, often overlapping geographically with other species of Kestrel, whilst in the Indian Ocean there are other species in the Seychelles *Falco araea*, in Madagascar *F. newtoni*, and in Mauritius *F. punctatus*; and probably, as Tristram points out, Réunion and Rodriguez had their own, now extinct, species.

Of particular interest to Tristram and included in his first collection was a series of pigeons of the genus *Alectoenas*. In the Seychelles *A. pulcherimma* has the body, wings and tail of a rich metallic blue-purple, the neck ash-grey and the crest red; Comoro *A. sganzini* is similar but lacks the red head whilst in Madagascar there is no red head, the head and neck are whitish blue and the tail red. An extinct form from Mauritius, represented by skins in Paris and the Natural History Museum at Tring, Hertfordshire, had the body and neck of the first two but possessed both red head and red tail. According to Tristram there are records of there being similar pigeons on Réunion and

Rodriguez of which there is now no trace. Tristram possessed specimens of three of these, from Madagascar (2), Seychelles (4) and Comoro (2).

Also gone are the indigenous parrots of Réunion, two having lingered until the end of the eighteenth century. The only endemic remaining in Tristram's time was thought to be the Réunion Cuckoo-shrike *Coracina newtoni*. However, other discoveries have occurred since, so that we know now that there are two endemic petrels and four other land birds, a Stonechat, a Bulbul and two species of White-eye. In addition a form of Starling which probably became extinct only in Tristram's time is also known from skeletal material. Jules Verreaux claimed to have shot one of the last remaining individuals in 1832. Some sixteen skins of this species exist in museum collections and two additional specimens in spirit (Tristram, 1877).

In Tristram's time there were eight species which were recognised as endemics to Mauritius: a Kestrel, a pigeon, a Parakeet, a Cuckoo-shrike, a bulbul, two White-eyes and a Tody. For at least two hundred years before Tristram wrote about oceanic islands, humans had altered the environment of many of those in the Indian Ocean. As a result it is likely that many more endemics existed until shortly before Tristram began his collections, and their demise is attributable mainly to human encroachment onto previously uninhabited islands. It was not just the people that caused the problems but the introduction of their satellites, i.e. cats, dogs, rats and particularly pigs which devastated ground-nesting birds. As early as 1710 Tristram cites a record of 1,500 introduced wild boars being killed in one day's hunt in Mauritius. What species were left alive of the previous endemics of the avifauna is listed in Table 7.

On the Seychelles fourteen endemic species remained at the time of Tristram's writing, nine on Mauritius, eight on Réunion, three on Rodriguez and at least five on Aldabra. On four islands a Tody was to be found; on three a Green Parakeet, a Bulbul and a Blue Pigeon were present; and on two a Kestrel, a Sunbird, a Drongo and a Cuckoo-Shrike. Now, fifteen endemics are recognised (Table 8).

Two extinct flightless birds (the Dodo and Solitaire), an unknown number of extinct flightless rails, and one extant flightless rail seemed to determine Tristram's approach to the birds of the western Indian Ocean. These, coupled with the stories and drawings of several fictitious flightless birds, seem to have coloured his view of the avifauna of the Mascarenes. Initially he sums up the situation quite logically: "Thus by the early [human] settlement of the Mascarene Islands we have lost many species, more perhaps than we dream of, and it is only within the last decade that naturalists have become aware of the importance of those insular forms and their bearing on the question of the distribution of species." Here, in his attempted explanation of the observable facts, Tristram suggests that "these insular species [and here he is referring only to the flightless birds, and the large ones at that, the Dodo and the Solitaire] have no direct descent from the fauna either of Africa on the one side or India on the other."

Tristram here is not his usual precise self, but if he means that it is not obvious that there are relationships with either African or Indian birds then he cannot be contradicted, but it is not clear. It is possible to see the direction in which Tristram is thinking when he writes: "We should be inclined to believe that these creatures were the descendants of progenitors other than those of the continents on either side, for their development was more feeble, their power of adaptation to circumstances infinitely less." Again there is lack of clarity. "We should be inclined to believe . . ." but does he really mean this? Is this Tristram not wanting to commit his real thoughts to print, is he uncertain of his position in his own mind or does he think that the readers of *Good Words* are not yet ready for the ideas of Darwin and Wallace? Perhaps *Good Words* was the correct place for Tristram to put forward his views on oceanic islands for this was an organ not only for story-tellers but also for more general articles. Tristram's rather vague attempts at an analysis of the facts come only at the end of Part III of 'The Story of the Isles of the Sea', and there are five more Parts to come. Fast-forward to the Presidential Address to the Biological Section of the BA in 1893: "It is only since we learned from Darwin and Wallace the power of isolation in the differentiation of species, that special attention has been paid to the peculiarity of insular forms" (Tristram, 1893)—Tristram's views put clearly!

## The Pacific Ocean: the Galapagos Islands

Tristram's consideration of the islands of the Pacific begins, not surprisingly, with the Galapagos Islands. He never visited them and his description of them is taken directly from Darwin's observations. These paint a sad picture of "dreary and unattractive cinder heaps". Darwin was not an ornithologist in the Tristram mould. Like the present writer, the latter would have regarded the islands, as he did the Farnes, as a magic place. He would also have appreciated the geology, which to say the least is impressive, and he would certainly have labelled the specimens he collected more accurately and usefully.

In Tristram's time the islands had not been so affected by humans as they have been more recently. As in the Mascarenes, pigs were released and, whilst the giant tortoises had been exterminated on some of the islands by seamen landing and seeking food, the birds were relatively unaffected: "The seamen had probably confined their hunting to the great tortoises, and, unlike the French sportsmen, had scorned to notice sparrows (finches) or thrushes (mocking birds)." Tristram suggested incorrectly that seaman having collected the great tortoises had transferred them to other parts of the world but elsewhere they had evolved completely separately. Like those of the Mascarenes the birds were very tame but they did not exhibit the more obvious morphological characters of the birds of the Indian Ocean. There were no great flightless birds—in fact the Galapagos Penguin and the Flightless Cormorant, which Darwin did not see, are the only flightless birds—and the only endemic rail is a very small bird, some fifteen centimetres long.

At Tristram's time of writing, he was in possession of Salvin's recent paper on 'The Avifauna of the Galapagos Archipelago' (Salvin, 1876) which provided the most up-to-date account available. Despite the fact that Tristram was writing almost forty years after Darwin's return from the Galapagos, several species of the extant birds had not then been described from the archipelago and the famous finches were yet still to be described and classified properly. Table 9 lists what Tristram thought to be endemic species, based on Salvin's work. He listed some thirty-three species of birds which were considered to be endemics, but accidentally omitted the Galapagos

Hawk which would make a total of thirty-four. Of these, nineteen were considered to be species of Darwin's finches but further work has shown several of these to be synonyms, three of which were shown to be synonyms of the Medium Ground Finch *Geospiza fortis,* two of the Cactus Finch *G. scandens,* two of the Vegetarian Finch *Platyspiza crassirostris* (Plate 63), two of the Large Tree Finch *Camarhynchus psittacula*, and a total of nine were wrongly identified. To these four new species were added, giving rise to the fourteen endemic Darwin's Finches now recognised. Concerning this Tristram was very far-sighted: "Some of these may perhaps blend into each other or be the peculiar races of particular islands. But the examination has been so hasty, that it is very probable other species still remain unnoticed." One of these was the Woodpecker Finch *Camarhynchus pallidus,* one of the few species of birds which uses a tool in the form of a twig, which is used to disturb prey in holes or hollow branches (Plate 64).

A great deal of work has been done on this group of finches since John Gould first demonstrated their relationships (Gould, 1837). Much confusion was caused by hybridisation between different species; and the change, over time, of the bill shape of certain populations of the same species, resulting from climatically-affected seed size and hardness (a short-term example of natural selection), has shown how bill shape alone was not a definitive character to be used in classification.

Tristram, speculating on the origin of the Galapagos birds points to continental America: "All the fifty seven species found here can be classified under thirty nine genera, of which twenty seven are world-wide, five are peculiar to the Galapagos and seven belong also to continental America." He goes on to write: "There is a remarkable absence of anything suggestive of a relationship with the rest of the Pacific. There is no honeybird, no flower-pecker, no shrike, no kingfisher, no parrot." He continues in comparing the Galapagos birds with the avifauna of South America: "no humming birds, though they occur on other islands nearer the coast, no tanagers, no tree-creepers (*Dendrocolaptes*) or Ant-hunters (*Formicaria*), of the ornithological features of South America, not a specimen is to be found. There is little colour in their plumage and no large (giant) pigeons, parrots and rails as in the Mascarenes." Again Tristram points out that they cannot be relics of a submerged continent as the islands are too young: "They are not Polynesian, they can only be American—but not South American." Tristram concludes that the Galapagos avifauna originated in North America. How far-sighted was this when over a hundred years later, with techniques never envisaged by Tristram, has work on the DNA of these birds suggested a Caribbean origin?

Sato, collaborating with other workers including Peter and Rosemary Grant (Sato et al., 1999), examined the mitochondrial DNA of thirteen species of Darwin's Finches and compared them with twenty-eight species of the Family Fringillidae (true Finches) and two outgroup species. The Dull-coloured Grassquit *Tiaris obscura* was found to be the nearest living relative of Darwin's Finches amongst the species examined. Adaptive radiation from the *Tiaris* group was found to have started in the Caribbean islands and spread to Central and South America, then to the Galapagos some 2.5 million years ago, around the time of the advent of the Pleistocene glaciations.

It cannot be claimed for Tristram that the idea was entirely his own. Darwin related the fauna of the Galapagos with that of the American continent and Salvin wrote: "The birds that are now found (in the Galapagos), being related to American birds, must have emigrated thence and become modified by the different circumstances with which they became surrounded." It appears from Tristram's writing that he too adopts the position taken by Salvin in relation to Darwin's Finches (Salvin, 1876). Tristram considered that "all the genera are strangely alike . . . of these genera there are known nineteen species [now only fourteen are recognised]. Some of these perhaps blend into each other, or be peculiar [belong exclusively to] races of particular islands." Here Tristram is not merely considering the possible evolution of a single species but of many species in similar circumstances, these evolving—though he does not use the word at this stage—into new species which diversify morphologically and are placed in different higher taxons, e.g., genera, admittedly artificially, but nevertheless, to some extent, indicative of further recognisable diversification.

## The Pacific Ocean: the Hawaiian Islands (Sandwich Islands)

"The birds of the Sandwich Islands are all resplendent with the brightest plumage"—so begins Tristram's account of them (Tristram, 1877). He compares them with the much more sombre Galapagos birds and speculates on their possibly different origins. Whilst the Galapagos finches almost certainly have an American origin, he concludes that in their origins the Sandwich Islands' birds are "as undoubtedly Polynesian as the Galapagos are American". He considered that another reason for the differences between these island groups lies in the difference of their geological ages and he rightly commented on this. He also drew attention to the fact that whereas the Galapagos were only recently colonised by humans it is now thought that this probably occurred over a thousand years earlier in the Sandwich Islands. The obvious result of this would be the longer effect of humans on the indigenous populations of organisms on the islands, and there appear to have been many more extinctions, even within living memory, than in the Galapagos.

Tristram was practically the first to study birds in many of the Pacific islands (Bodenheimer, 1957a) but he did this through his collections, which he acquired through purchases, exchanges and the collections made by his missionary friends. As a result of this some islands were not as well represented as others in his collections, and this was the case for the Sandwich Islands from where he possessed only four species of Hawaiian Creepers, the O'u, O'ahu Akepa, I'ive and Agapane in his first collection and none in the second.

The work upon which Tristram based his comments on the general avifauna of the Sandwich Islands was that of Sclater (1871) and he referred in the main to the endemic species. Of "a peculiar crow" he says that "this species is restricted to these islands" and "he has been long enough domiciled to assume distinctive, if only slightly distinctive, characters." Clearly Tristram is suggesting here that it has become distinctive (to specific level) because of its isolation. Tristram follows Sclater closely in his recognition of two "Flycatchers", belonging to "families entirely unknown on the American Continent" but "yet nearest in character to the flycatchers of Polynesia." The other endemic forms of land birds belonging to the group referred to by Tristram as "Honeysuckers" are now usually referred to as Honeycreepers. The group consists of individual species that have evolved into nectar-feeding, probing, seed-eating, nut-cracking and parrot-billed forms, and one even occupies the woodpecker niche. The three species of Moho, together with a fourth described after the publication of Tristram's paper, are all now extinct, due largely to the Hawaiian natives killing them for their feathers (Table 10). Tristram writes that "under the wing it has three or four yellow feathers; these were collected to form a kingly mantle, which demanded many thousand Mohos for its completion, which is said could not be accomplished under three generations, and when finished was priceless." He further wrote: "Alas, the introduction of firearms has all but completed his extermination and it is now nearly impossible to obtain a specimen." Tristram did not acquire one!

It was not only the Mohos that were collected by the Hawaiian natives; some of the smaller "sickle birds", particularly those with red feathers, were slaughtered to decorate the helmets of the warriors. The term "sickle bill" is applied to this group because of the remarkable shape of the bill, which is in some cases semi-circular. Different species have the curve adapted for taking nectar from variously shaped flowers or for gathering insects. The variety of bill is much greater than in the Galapagos finches but similar to these birds in that they are each thought to have evolved from a common stock which reached each group of islands at an early stage of their existence and rapidly evolved into different ecological niches (adaptive radiation).

The original ancestor of the honeycreepers is thought to be a bunting or finch-like bird which originated from the continent of America (Fisher and Petersen, 1964). Tristram thought that "what affinities they have are Polynesian, however distant", but more recent work (Lerner et al., 2011) has shown that the most likely ancestors are rose-finches of central Asian origin. Rose-finches are an eruptive species and the DNA analysis carried out suggests a date of between 5.8 and 7.2 million years BP, some two million years later than the colonisation of the Seychelles by other birds.

The Sandwich Islands, or the Hawaiian Islands as they are now known, had some seventy-one species or sub-species of birds which can be regarded as endemics; twenty-three of these are now extinct, and of the remaining forty-eight, thirty-two are regarded as endangered. A significant number of these have been described since Tristram's time and others have become extinct. As on other oceanic islands human beings themselves and introductions by humans have been mainly responsible for this loss.

Having outlined the very specific avifauna of the Galapagos, the Seychelles and the Sandwich Islands, Tristram found from the study of the skins sent to him that "the bird families of all the other Pacific groups (of islands) . . . must, without any hesitation, be classed together as one region of life." All the islands are of volcanic origin and Tristram held that "whatever be the origin of the bird life of Polynesia, it certainly shows few of the peculiarities which stamp the Mascarenes or New Zealand with the impress of geological antiquity. The largest and highest islands, those of Fiji, Tonga and Samoa are centrally situated." Westward, Tristram records islands far richer and more varied in bird life (the Solomon Islands and the New Hebrides) and attributed this to the proximity of Australia and New Guinea, whilst eastwards, towards the Marquesas and Society Islands, the number of species grows fewer. From Tristram's writings it is clear that he did not include such islands as Madagascar and the Philippine Islands in his definition of oceanic islands, which for the most part were away from the nearest continental shelves and of volcanic origin.

In considering this series of islands, Tristram remarked on the brilliance and contrast of colours of the parrots and pigeons "of each of which almost every island possesses several species". Tristram was particularly interested in species which were confined to very limited areas and one in particular attracted him. This was the Tooth-billed Pigeon of Samoa, of which he possessed three skins and a single specimen in alcohol in his first collection and a single skin in his second collection. *Didunculus stigirostris* or Little Dodo is still to be found in Upolu, though in Tristram's time it was very rare and originally known only from a single specimen in the collection of Sir William Jardine and another "obtained by the United States Exploring Expedition which was lost by shipwreck". Once common, it was almost exterminated by cats released by voyagers, as apparently it bred and roosted on the ground, or at least some did so. They survived by some taking to the trees and nesting there, as probably some had previously done, and now there is a population of some five to seven thousand.

Another bird which interested Tristram was *Lamprolia;* now it is known as the Silktail. He speculated that it might be distantly related to the Rifle-bird of Australia. Subsequently different workers have attributed it to various other genera but in 2007 the problem was solved by DNA sequencing (Irestedt, M. et al., 2008). In this work a sister relationship between the Silkbird and the Papuan Mountain Drongo was supported and these in turn to the fantails (*Rhipidura*), a distance of over 3,000 miles (4,830 km) separating the present populations. Distant though this is, other work has recently shown that this is a possibility. There are three possible scenarios, two involve a long-distance dispersal either by the ancestor of *Lamprolia* to Fiji or the ancestor of the Drongo to Papua New Guinea. The third possibility involves the utilisation of island areas at subduction zones where islands are formed and submerged continuously (at tectonic plate margins). Had the resources used in this work been available in Tristram's time there can be little doubt that he would, with the help of Newton, have found someone to carry out the work, and more like it. He would have been delighted with the results.

Much of the material which Tristram studied from the Pacific was provided by E. L. Layard who, at the time, was working as a civil servant in Fiji and New Caledonia. He sent skins to Tristram, who identified and described them, many of them new species, the descriptions of which were subsequently published in the *Ibis*. Layard carried out a lot of field work on different islands on which he was collecting and showed that "almost every island of each cluster has its own particular parrot and pigeon affined, it is true, to their nearest neighbours, but yet permanent and unvarying in their characteristic features." The most conspicuous and abundant birds in the region were the pigeons, represented in Tristram's time by five genera and now by more than double that number. Tristram mentions "*Didunculus* in Samoa, *Carpophaga* the colourful tree doves, *Ptilinopus* everywhere and turtle doves and metallic pigeons *Janthaenas* in most of the islands. Most are confined to a single group of islands and many of the tree doves to a single islet." Tristram observed that they differed from all other pigeons in that they are sexually dimorphic, the males being very colourful and the females being "a nearly uniform pale green".

Many of the Polynesian species that Tristram examined he found to resemble Australian species of the same family "but not one is identical with any known beyond the Polynesian region". He also observed that in eastward progression through the Polynesian region the colouring becomes less brilliant. This is particularly so in the small parakeets which Tristram included in the Genus *Coriphilus* which is no longer recognised. These are now included in the genera *Phigys*, *Charmosyna*, *Glossopsitta* and *Vini*. The birds from the islands nearest Australia have a plumage of brilliant yellow, red, blue and green, with usually some blue on the head. Blue then dominates towards the east and in the Society Islands there is a parakeet which is completely bluish-purple. Again, the pigeons similarly have less colour in their plumage in the eastern Pacific islands. Tristram also noticed that the strongly marked species are to be found only in the forests of the bigger islands, "never on the atolls of the coralline groups".

Nowhere else in the Pacific islands did Tristram observe "the weakness of flight" indicated by shorter wings or the wings being reduced to the extent that the birds were flightless, as in the Mascarenes: "The birds of Polynesia seem to have been providentially adapted for flying themselves over a vast insular area." This is probably the first time that Tristram used the word "adapted" in print (Tristram, 1877).

## The Pacific Ocean: New Zealand

As with those he encountered in the Mascarenes, Tristram was fascinated by the flightless birds of New Zealand and the evidence which existed of former giant forms of such birds. He pictured *Dinornis* as a defenceless creature (for what was there to defend against?) feeding on foliage and roots, being hunted not only by the first humans in New Zealand but by the huge *Harpagonis*, a colossal bird of prey for which there was sub-fossil evidence. Co-existent with this were Gigantic waterhens (*Aptornis* and *Notornis*) present in the marshes, and a giant goose (*Cnemiornis*) could be found in the lagoons. Whilst we have the arrival of humans in New Zealand to thank for the loss of these great birds we have to thank them also for the preservation of their bones so that we may know what they were like. It was Tristram's colleagues and contemporaries who reconstructed their one-time appearance from their remains. These remains were not true fossils but what could be found in the ground ovens which were used to cook these birds as food by the native populations of the islands. These ovens, as Tristram writes, were the equivalent of the kitchen middens of northern Europe, and if there was something there to be collected from them, Tristram could be relied on to collect it—or he could rely on the missionary zeal of those whom he had sent out to collect it!

In his 'Story of the Isles of the Sea' he wrote: "I lately received from one oven a number of thigh, leg and toe bones, and the pelvis of one of the smaller species of *Dinornis* with the tarsus and other fragments of *Apterix* or Kiwi, and portions of at least three young children, probably all part of the same feast." There is no record of that which was received ever being reported to the senders, or to Mrs Tristram, but they could have read about it later in *Good Words*.

Tristram speculated that it might be possible to arrive at a date for the separation of the two islands of New Zealand. Each island possesses closely related but distinct species so making it theoretically possible in Tristram's time but not practically so until more recently. Northern and Southern Kiwis and Weka Rails can be so distinguished, as can the choughs, thrushes, warblers, wrens, titmice and stonechats. As Tristram wrote: "It is in all these cases impossible to doubt the common ancestry, but the disruption must be very ancient to have produced these permanent modifications through almost the whole range of bird life."

For some time during his career Tristram was of the opinion that most of the large, flightless birds were related and had a common origin and he was also interested in the time of their extinction. At this stage in the development of ornithology it was impossible even to estimate dates except in a few cases. Captain Cook's expedition saw nothing of the Moas. However,

York museum still possesses a skeleton of the South Island Moa *Dinornis robustus* (Plate 65) to which Tristram referred. The specimen was found by gold prospectors in a sand drift and had borne fragments of skin and feathers. Large fragments of eggs were also found and it would appear that the bird had been overwhelmed in some way, possibly by a sand storm, and buried at its nest. Possibly a clearer record of the co-existence of human beings and Moa was that of the skeleton of a Maori who, clutching the egg of a moa, was found in a sitting position in a cave. This might suggest that the two were contemporary. Tristram also records that moa hunting featured in Maori sounds and that the native chief Rauparaha was buried in 1849 together with a moa feather, apparently the last relic of the moa still held by the tribe.

Tristram mentions eleven species of moa recognised in his time but now only nine species are accepted. These vary between 5 feet (153 cm) and 12 feet (366 cm) in height and were apparently all contemporary as they were found in the same alluvial deposits. Here too was found *Harpagornis* which Tristram states was so large as to be able to carry a Moa or a human. Whilst he was particularly interested in these extinct forms he was also enthralled by the living birds which, again, he found to be very different from those of the rest of the world. Of those that came to his notice (there were others yet to be described) he found that "no other region contains so large a proportion of peculiar genera". Here he is using the adjective to describe this singular attachment to the islands rather than describing the bird as an oddity. Of the whole 145 native land birds, he observed that only three were identical with other birds of the southern hemisphere, two of the three being migrant cuckoos and the third an immigrant white-eye from Australia, which arrived and established itself within the memory of living settlers.

Since the start of the colony the introduction of cats, dogs and pigs adversely affected many species. The release into the wild of species of English birds seemed to have the same effect and the introduction of honey bees (there was no New Zealand honey bee) which quickly established themselves proved to be catastrophic for the Bell-birds which were stung when competing with the bees for nectar. Fruit-eating and insectivorous birds were adversely affected by the introduced rats which spread quickly, particularly affecting hole-nesting birds.

Tristram made a lot of his comparisons using genera (artificial groupings of species): "the 60 or 70 species of land birds come under 34 genera of which 16 are peculiar to New Zealand. Five genera are also found in Australia but all the resident species are distinct." Because of changes made to genera and species over the years, mainly related to the establishment of the International Rules of Zoological Nomenclature, this is difficult to translate into real numbers. However, it has been estimated that some 115 endemic species existed in New Zealand at the time of the first human occupation and now nearly 40 per cent of these have been lost. There are now about seventy endemic species alive, many of which are endangered. All the species of moa are extinct and nine of the thirteen rails. Nearly 140 years ago Tristram picked out a number of interesting species of birds to illustrate their plight in New Zealand. The Laughing Owl preyed largely on the only indigenous mammal, the New Zealand Rat, but this was already extinct as a result of competition from the introduced Norway Rat which was too large for the owl to take as food; there has been no substantiated record of the owl since 1914. The owl-like Kakapo, a flightless ground parrot even though it possesses fully developed wings, can glide, however, for up to 100 m from a high perch. Known as the owl-parrot because it possesses an owl-like facial disc of feathers, it was rare in Tristram's time and he thought it on the edge of extinction. He had two skins of the Kakapo in his collection. Somehow it has survived until today though it is now considered to be critically endangered. A third parrot selected by Tristram for comment was the Kaka, a fruit-, insect- and nectar-eating bird, with an overlapping upper mandible. Dark coloured, in contrast with most of the parrots of New Zealand, it has also managed to hold its own but is still regarded as endangered.

Turning to the Passerines or Singing birds, Tristram possessed five specimens of the Huia in his first collection, a bird the size of a Jackdaw, with glossy black plumage and orange wattles at the sides of its bill. Now extinct, the male possessed a stout, curved bill which it used like a woodpecker whilst that of the female was, unusually, much longer, finer, sickle-shaped and used for probing. According to Tristram's correspondent, Sir Walter L. Buller, the birds fed co-operatively, the male drilling decayed wood for insect larvae, the female often retrieving the prey with her finer bill (Tristram, 1877).

In the nineteenth century there was only one pigeon in New Zealand, the New Zealand Pigeon, and this was a large-sized fruit eater. Subsequently three more species were introduced, but the single endemic species in the New Zealand landmass was in marked contrast with the rest of the southern hemisphere where a variety of pigeons was a common feature of most islands. Similarly there was only a single species of gallinaceous bird, the New Zealand Quail, now extinct. New Zealand certainly possessed more than its fair share of curious birds and the Wrybill was and is one of the most curious. Its bill is turned to the right and it scoops sideways to pick up most items of prey.

Perhaps the most widely known bird of New Zealand is the Kiwi, of which Tristram claimed there to be four species, whereas only three, the Brown, the Little-spotted and the Great-spotted were recognised until 1993. Then the Okarito Kiwi *Apteryx rowi,* was separated from the Southern Brown Kiwi, on the basis of dissimilar DNA, and *A. mantilli* was separated in 2000, so that at present five species are recognised. (Shepherd and Lambert, 2006). Tristram drew attention to the fact that the Kiwi's egg is larger than that of any other bird in relation to its body size: "the egg of a mother weighing three pounds (1364 g) will weigh fifteen and a half ounces (440 g) when filled with water."

Tristram was pessimistic about the future of the avifauna of New Zealand about which he stated that it was "entitled to claim for itself a genealogy and an origin widely removed from that of any other region." Yet he accepted "more than half of her birds are gone"—extinct! "In a few years the omens are too clear, we shall have to look for the story of New Zealand, not in her swamps and fern brakes, but only in the gravel pits of her rivers and the museums of Europe." Active conservation measures have been taken to preserve endemic species in New Zealand, for example the Takahe *Porphyrio mantelli* (Plate 66), which was thought to be extinct in 1948 until a few were found in Fjordland and now still survive in the wild and in sanctuaries such as Tiritiri Matangi.

Surrounding New Zealand are many other islands which once had their own unique birds. Lord Howe Island, Norfolk and Philip Islands, the Kermadec group to the north and west and the Chatham, Auckland and Macquaries groups to the south and east. These islands all possessed birds which were not Australian or Polynesian but related to those in New Zealand, which we now know to have been separated from any other land mass since the time of the dinosaurs, some 65 million years ago. Tristram did not know this nor did he know accurately the age of the different volcanic islands that rose over time from the sea bed of the Pacific, but he did know that there was a great deal of difference between the bird life of the two islands of New Zealand.

## The Pacific Ocean: Australia

The avifauna of Australia was not particularly well documented in Tristram's time and he refers to 630 species of which "not above one thirtieth occur elsewhere and everything that specifically characterizes the neighbouring Indian region is absent." Now nearly 850 species are recognised but it remains true that Australia has its own particular bird fauna. In his review he highlights what is not present: "no pheasants, no woodpeckers, no tree finches and most raptors are exclusively Australian." Grass finches replace the true tree finches with the exception of the introduced European Goldfinch and Greenfinch, and kingfishers are abundant and varied. The Bower birds and Rifle birds have moved in from Indonesia and the Australian Bustard is very similar to its relatives in India and Africa and, for Tristram, "he has no business here".

As in Polynesia, parrots and pigeons provide a variety and richness of plumage and, of the land birds, form a large proportion of the whole. Tristram pointed out that the pigeons are not so varied as in Polynesia but have a relationship with them greater than with those of Indonesia. Tristram picked out the parrots for special attention and particularly the Brush-tongued Lorikeets whose tongue is specially modified to extract nectar from flowers. An unrelated family, but one occupying much the same geographical areas, is that of the Honeycreepers; these, too, have a brush-tongue. Tristram points out that Alfred Russell Wallace observed that "the presence or absence of these brush-tongued birds serves to define and limit the Australian region with a presence hardly equalled by any other family of birds."

The Brush Turkeys (Mound Builders) took Tristram's attention because of the unique method of incubation. The adult birds, both of a pair, build a mound of leaves and earth up to five feet (ca. 150 cm) high and four to five feet (ca. 120–150 cm) in circumference, in which the female lays her eggs. They are arranged in a circle, on the ends, in the manner of some dinosaurs, and incubated by the rising temperature of fermentation within the mound. Incubation lasts for about seven weeks and the young dig themselves out of the mound on hatching. They are immediately able to run and flutter up into trees. Mound birds are found in drought areas where there is often a shortage of food and the method of incubation allows the adults to roam further afield as they do not need to attend the nest.

Australia has many more interesting birds but Tristram set out to write about the islands of the sea and, whilst certainly it is an island, Australia is also a continent and is best treated as such.

## The Pacific Ocean: Papua New Guinea

Despite having good collections of bird skins, including thirty-six birds in total in the two islands, Tristram chose not to comment at length on the avifauna of Indonesia. However, he was much attracted to the Kingfishers, particularly the "racket-tails" (*Tanisiptera*) and the Birds of Paradise. He comments that "until recently no European had been able to observe them in life [and] all our specimens were supplied by the natives who always cut off the legs from the skins, on which account they were reputed to be without feet, whence the name of the best known species"—the Greater Bird of Paradise *Paradisaea apoda*. They consist of a group which Tristram suggested vary so widely that many other systematists would have given generic distinction to all species, so indicating an origin very distant to New Guinea. There are no links with the forms of any other island, except the Rifle Birds of Australia.

Due to his particular interest in flightless birds Tristram could not overlook the existence of the Cassowaries in Papua New Guinea and quoted Newton (1896) as recognising nine different species. Newton based his comments on the work of Salvadori (1880–2) as the authority and who even suggests a tenth species. Modern taxonomy recognises only three but needs revision as some systematists recognise up to nineteen sub-species. Of the three recognised species the Northern Cassowary is limited to northern New Guinea and neighbouring islands, the Southern Cassowary to most of New Guinea and extends into northern Queensland and Seran, and the Dwarf Cassowary is found in New Guinea and Japan Island off its north coast.

Tristram ends his review of oceanic islands with a very brief survey of the islands close to China and India and despite there being many endemic species associated with them refers them to the adjacent continents. Similarly he allocates the Atlantic Islands of the West Indies to the adjacent land mass of the Americas and points to the fact that of 350 species in Trinidad and Tobago "not one is peculiar to these islands and are simply South American". In making this statement he explains it on the basis of "a very shallow sea" separating the islands from the continental land mass and draws attention to the distance between these in comparison with the much greater distances of oceanic islands from their adjacent continents, thus associating the greater separation with endemism without actually stating it.

Nowhere at this time—1877—except in the vaguest of terms, does Tristram consider the implication of his observations. In his papers in *Ibis* and the Proceedings of the Zoological Society he remained descriptive and in identifying birds collected by others his reports were similar. To the outside world he remained a field naturalist, "the helpful servant of the philosopher and systematist". But he was much more than that. He was a very capable systematist to whom others sent their collections for his comments, both oral and written, and he was also a very able philosopher who was capable of keeping much of his thinking to himself but perhaps not from Alfred Newton. Newton being Newton undoubtedly chided him in private about his position vis-à-vis Darwinism, but appreciated the position taken by his friend. Tristram thought about oceanic islands and their association with endemics, and about the comparison between the avifaunas of islands and large land masses. He thought about those birds that had lost the power of flight and particularly the circumstances in which they flourished or became extinct and he discussed all these matters with his close associates.

However, the readers of *Good Words* were left to draw their own conclusions from Tristram's account of oceanic islands and probably few appreciated the implications of what had been written and what they had read. The eighth and last part of the story ended rather abruptly with a quick review of islands in the Indian and Pacific Oceans that had not been previously mentioned, but there was no summary or conclusions.

During the course of the work several points were made clear: firstly, that flightless birds of several different ancestries have appeared on a widely dispersed variety of oceanic islands; secondly, several species of birds which occurred on oceanic islands were found to be closely similar to forms on neighbouring islands, but in many cases were specifically different; thirdly, the absence of predators on oceanic islands had resulted in many of these forms thriving whereas as soon as predators were introduced their existence was threatened and in many cases extinctions occurred; fourthly, many of the forms thrived in isolation and it was tempting to think that this might have contributed in some way to changes taking place leading to specific differentiation. All these matters must have been discussed at length within the developing group of ornithologists which formed the expanding BOU, and much of this discussion was led by Alfred Newton, now a firm supporter of Darwinism. Darwin himself was not an ornithologist but his co-author of the original Linnean Society paper, Alfred Russell Wallace was, and also an active member of the Union and a contributor to the *Ibis*. The original group of founding members, with the exception of the deceased John Wolley, was still very active and their views must have caused Tristram to reconsider any lingering doubts that he had relating to the views of Darwin and Wallace.

The support for Darwinism was increasing, not only in the academic world but amongst the more progressive elements in the church and in the world at large as time went by. However, whatever his private thoughts were or what he said within the inner circle of the British Ornithologists' Union, it was not until sixteen years after the publication of 'The Story of the Islands of the Sea' that Tristram committed himself to a scientific audience and to the published record of the BA meeting of 1893.

Convinced as he was at an early stage of his explorations of the powers of natural selection within the species, his consideration of the distribution of what were obviously closely related species occupying separate islands, must have gone a long way to convincing him that species do change. His examination of islands such as the Galapagos and many others in the Pacific was indicative of this having happened. He had seen for himself that species were not immutable and in having some concept of the age of the various islands, which he frequently referred to in his writings, this was an indication of how long the process of change might take. Tristram must have given a lot of consideration to his observations but did not commit his conclusions to paper nor publish any conclusions to which he came in any scientific journal. In what was still a contentious area of natural history it is probable that he wanted to maintain his conservative position in the church so that he deliberately avoided the continuing debate on evolution, particularly in print. In private there can be little doubt that he was following Newton and his other contemporaries towards the quickly changing climate of committed Darwinism.

Nowadays the study of genetics has provided evidence which was not available to Tristram and his friends, and it is generally acknowledged that evolutionary theory is based on four main contributory sources, mutation, recombination, selection and isolation, originally proposed by Ernst Mayr (1966). Tristram knew nothing of the first two of these pillars but he did know about selection and isolation and certainly had available to him all that was then known of these topics. It was enough for Newton to claim immediately after the Oxford debate, in a letter to Tristram: "I am quite converted. I was (I confess it) in a 'state of transition' but Darwin*oid* I might have remained for a whole geological aeon. The Bishop's speech and article have caused me by a process of 'natural selection' to become something better. I am developed into pure and unmitigated Darwinism." (Letter 9839/1T/209)

Thirty-two years later, in his Presidential Address to the Biological Section of the BA (1893) in Nottingham, read by Sir W. H. Flower because of Tristram's illness, it was stated: "It is difficult for the mind to grasp the advance in biological science which has taken place since I first attended the meeting of the British Association some forty years ago. In those days, the now familiar expressions of 'natural selection', 'isolation', 'the struggle for existence', 'the survival of the fittest', were unheard of and unknown." By this time Tristram's interests had turned to the avifauna of islands, particularly oceanic islands, and later in his address he wrote of how "It is only since we learned from Darwin and Wallace the power of isolation in the differentiation of species, that special attention has been paid to the peculiarities of insular forms. Here the field naturalist comes in as the helpful servant of the philosopher and the systematist, by illustrating the operation of isolation in the differentiation of species."

Then in his address Tristram turned his attention to his desert larks and repeated his 1859 stance: "It is hardly possible to illustrate this theory [natural selection] better than by the larks and chats of North Africa" and gave further examples including the geographical changes in the morphology of Blue Titmice. In illustrating the importance and value of field observations Tristram considered aspects of migration, mimicry, parasitism, heredity, bird architecture, embryology and several smaller fields of interest, all of which were then in their infancy, and thus demonstrated how much was yet to be found. He ended his address by referring to "our great master Darwin" but as a churchman could not conclude without resort to an attribution to an ultimate creator. So Tristram was a churchman but he was also a Darwinian and who can say that it is not possible to be both?

Eleanor Fleming (n.d.), a grand-daughter of Tristram, wrote a very comprehensive appreciation of her grandparents, of whom she was obviously very fond. Though this was never published, several copies exist in the possession of the Canon's descendant relatives and it contains numerous comments on her grandfather's conservatism. One in particular is pertinent to Tristram's position on natural selection:

> In many ways grandfather was very conservative in his views and although a great admirer of Bishop Westcott, then the Bishop of Durham who succeeded Bishop Lightfoot, both great scholars who translated and produced with Hurst the revised version from the Greek New Testament, yet grandfather always spoke of it as the 'Reversed Vision', and of Hymns Ancient and Modern as 'Hymns Pagan, Popish and Protestant'. I think he believed in the Verbal Inspiration of the Bible, and hated Bible criticism as being disloyal; but he went 'all the way' with Darwin and said to me once 'When the world was evolved, oh! Created . . .'.

"Evolved" is not a word that Tristram often used and certainly not in print. He and Newton were such close friends that it is likely that Newton supported Tristram in his avoidance of any form of conflict, either orally or in print, concerning his views on Darwinism. It is unlikely that their views differed greatly on that subject, and as Wollaston points out, Newton still attended church on Sundays as there were still some aspects of the old religion that Newton was prepared to accept. He understood that Tristram, as a churchman, must be seen to support his peers, at least in public, and that his private views should be kept private whilst controversy lasted. "The meeting [of the British Association] at Cambridge in 1862 witnessed the last determined resistance of the anti-Darwinians and their ultimate defeat." (Wollaston, 1921) In the outside world acceptance of Darwinism was to take longer. As the climate changed over the years so it was easier for Tristram to express his views, and this he did. They were no secret to his family, as Eleanor's reflections make clear, and he was less concerned with the opinions of others as time went by. In October 1867 Tristram wrote to Newton: "I hope you saw the 'Pall Mall' on Friday on my paper at [Wolverhampton?] . . . I spoke out much more manfully in defence of your Darwinism than is shown in the papers and certainly carried the meeting." (NCP 9839/IT/227)

Another of Eleanor's reflections is of interest in this context: "Bishop Walpole [sometime head of the Church of England Training College in Durham] . . . when he was once enlarging on grandfather's wide interests and great gifts in which he was an authority, to which I answered 'Yes, on every subject except his profession as a clergyman'. He laughed and said 'You are quite right.'" Years later, of F. C. R. Jourdain, his daughter wrote similarly but less kindly that he "sacrificed his family and also his priesthood to ornithology" (Spurling, H., 1995)

By coincidence, (or was it?) on the day that Darwin first wrote to Tristram, 4 June 1868, Tristram was elected to the Royal Society. On 6 June 1868, in his reply, Tristram wrote: "I am glad to have the opportunity of thanking you for your unexpected kindness in putting your name to my certificate for the Royal Society." (Letter DCP6234) This was a recognition by his peers of his scientific achievements and of his development over the years as a thinker along scientific lines. The latter is clearly demonstrated by the marked contrast between his two Presidential Addresses, the first in 1860, the other in 1893. In his first he was an embryonic Darwinian, and at the time of his second a scientist who considered Darwinism to be the best theory put forward to that date to explain the phenomena of the biological world, as it is even now. Unlike Newton, he had not "developed into pure and unmitigated Darwinism" at a stroke, but he had certainly arrived at a firm acceptance of it. Canon Henry Baker Tristram was a Darwinian, and a great one at that! In the end, and it may have taken thirty years, Tristram was just as enthusiastic as Newton: "It is only since we learned from Darwin and Wallace the power of isolation in the differentiation of species that special attention has been paid to the peculiarities of insular forms" (Canon H. B. Tristram, Nottingham, 1893, Presidential Address, Biological Section of the British Association) and at that time no-one knew more about insular forms than he!

Further confirmation of Tristram's acceptance of Darwinism comes from Newton himself: "He [Tristram] had to modify his expressions sometime after [the Oxford 'Debate'], when the 'orthodox tide' was flowing, just as Galileo was obliged to do so, but he held them all the same to the end, and great credit is due to him for this." This appeared in the Royal Society Obituary for Tristram written by Gunther (1908) who quoted it as an MS note from Newton.

Table 7: Mascarene Islands: Endemic species of birds

| Mauritius endemics | Tristram's endemics | Present-day endemics |
|---|---|---|
| Mauritius Kestrel | *Falco punctatus* | *Falco punctatus* |
| Pink Pigeon | *Columba mayeri* | *Nesoenas mayeri* |
| Mauritius Parakeet | *Palaeornis echo* | *Psittacula echo* |
| Mauritius Cuckoo-Shrike | *Oxynotus typicus* | *Coracina typica* |
| Mauritius Bulbul | *Hypsipetes olivaceus* | *Hypsipetes olivaceus* |
| Mauritius Olive White-eye | *Zosterops chloronothus* | *Zosterops chloronothus* |
| Mauritius Grey White-eye | *Zosterops modesta* | *Zosterops borbonicus **mauritianus*** |
| Mauritius Tody | *Foudia rubra* | *Foudra rubra* |
| Mauritius Blue Pigeon | *Alectroenas nitidissima* | EXTINCT |

| Reunion endemics | Tristram's endemics | Present-day endemics |
|---|---|---|
| Reunion Stonechat | | *Saxicola tectes* |
| Reunion Black Petrel | | *Pterodroma aterrima* |
| Barau's Petrel | | *Pterodroma baraui* |
| Reunion Cuckoo-Shrike | *Oynotus newtoni* | *Coracina newtoni* |
| Reunion Bulbul | | *Hypsipetes bourbonicus* |
| Reunion Olive White-eye | | *Zosterops olivaceus* |
| Reunion Grey White-eye | | *Zosterops b. **borbonicus*** |
| Bourbon Crested Starling | *Fregilupus varius* | EXTINCT |

| Rodrigues endemics | Tristram's endemics | Present-day endemics |
|---|---|---|
| Rodrigues Warbler | *Drymoeca rodericana* | *Acrocephalus rodericanus* |
| Rodrigues Fody | *Foudia flavicans* | *Foudia flavicans* |
| Rodrigues Green Parakeet | *Palaeornis exsul* EXTINCT | EXTINCT |

Table 8: Seychelles Endemics

| | Tristram's endemic | Present-day endemic |
|---|---|---|
| Seychelles Kestrel | *Tinnunculus gracilis* | *Falco araea* |
| Black Parrot | *Coracopsis barklyi* | *Coracopsis nigra barklyi* |
| Seychelles Green Parakeet | *Palaeornis wardi* | *Psittacula eupatria wardi* |
| Seychelles Sunbird | *Nectarinia dussumieri* | *Nectarinia dussumieri* |
| Seychelles (Grey) White-eye | *Zosterops modesta* | *Zosterops modesta* |
| Chestnut-flanked White-eye | *Zosterops mayottensis* | *Zosterops mayottensis* |
| Seychelles Black Paradise Flycatcher | *Tchitrea corvinus* | *Terpsiphone corvina* |
| Seychelles Magpie Robin | *Copsychus sechellarum* | *Copsychus sechellarum* |
| Seychelles Bulbul | *Hypsipetes crassirostris* | *Hypsipetes crassirostris* |
| Seychelles Tody | *Foudia sechellarum* | *Foudia sechellarum* |
| Seychelles Blue Pigeon | *Erythroena pulcherima* | *Alectoenas pulcherima* |
| Turtle Dove | *Turtur rostratus* | *Streptopelia picturata rostrata* |
| Seychelles Scops Owl | *Gymnocops insularis* | *Otus insularis* |
| Seychelles Swiftlet | *Collocalia fusifaga elaphra* | *Aerodramus elephras* |
| Aldabra Drongo | *Buchanga aldabrana* | *Dicrurus aldabranus* |

Table 9: Galapagos: Endemic species of birds

| Common Name | Scientific Name (Tristram's time) | Current Scientific Name |
|---|---|---|
| Galapagos Penguin | *Spheniscus mendiculus* | *Spheniscus mendiculus* |
| Waved Albatross | Salvin 1883* | *Diomedea exulans* |
| Wedge-rumped Petrel | *Procellaria tethys* | *Oceanodroma tethys* **tethys** |
| Dark-rumped Petrel | *Oestrelata phaeopygia* | *Pterodroma phaeopygia* ***phaeopygia*** |
| Flightless Cormorant | Rothschild 1898* | *Nannopterum harisi* |
| Lava Heron | *Buteroides plumbeus* | *Butorides sundevalli* |
| Yellow-crowned Night Heron | *Nycticorax pauper* | *Nyctanassa violacea* **pauper** |
| Galapagos Hawk | *Buteo galapagensis* | *Buteo galapagoensis* |
| Galapagos Crake | *Porzana spilanotus* #1 | *Laterallus spilonotus* |
| Lava Gull | *Larus fuliginosus* | *Larus fuliginosus* |
| Swallow-tailed Gull | *Creagrus furcatus* | *Larus furcatus* |
| Galapagos Dove | *Zenaida galapagensis* +1 | *Zenaida galapagoensis* |
| Galapagos Flycatcher | *Myiarchus magnirostris* #2 | *Myiarchus magnirostris* |
| Galapagos Mockingbird | *Mimus parvulus* | *Nesomimus parvulus* |
| Charles Mockingbird | *Mimus trifasciatus* +1 | *Nesomimus trifasciatus* |
| Hood Mockingbird | Ridgeway 1890* | *Nesomimus macdonaldi* |
| Chatham Mockingbird | *Mimus melanotus* | *Nesomimus melanotis* |
| Small Ground Finch | *Geospiza fuliginosa* #2 | *Geospiza fuliginosa* |
| Medium Ground Finch | *Geospiza fortis* #2 | *Geospiza fortis* |
|  | *Geospiza dubia* |  |
|  | *Geospiza dentirostris* |  |
|  | *Geospiza nebulosa* |  |
| Large Ground Finch | *Geospiza magnirostris* +2 | *Geospiza magnirostris* |
|  | *Geospiza strenua* #2 |  |
| Sharp-beaked Ground Finch | Sharpe 1888* | *Geospiza difficilis* |

| Common Name | Scientific Name (Tristram's time) | Current Scientific Name |
|---|---|---|
| **Cactus Finch** | *Cactornis scandens #2* | *Geospiza scandens* |
| | *Cactornis assimilis* | |
| | *Cactornis abigdoni* | |
| **Large Cactus Finch** | Ridgeway 1894* | *Geospiza conirostris* |
| **Vegetarian Finch** | *Camarhynchus crassirostris +2* | *Platyspiza ceassirostris* |
| | *Camarhynchus variegatus #1* | |
| | *Camarhynchus prosthemelas* | |
| **Small Tree Finch** | *Geospiza parvula #2 +1* | *Camarhynchus parvulus* |
| **Medium Tree Finch** | Ridgeway 1890* | *Camarhynchus pauper* |
| **Large Tree Finch** | *Camarhynchus psittacula* | *Camarhynchus psittacula* |
| | *Camarhynchus habeli* | |
| **Woodpecker Finch** | *Cactornis pallida* | *Cactospiza pallida* |
| **Mangrove Finch** | Snodgrass & Heller 1891* | *Cactospiza heliobates* |
| **Olivaceous Warbler Finch** | *Certhidea olivacea #1* | *Certhidia olivacea* |
| **Grey Warbler Finch** | *Certhidea fusca* | *Certhidia fusca* |

\* Authority of species named later than Tristram's paper
\# Tristram's First Collection, with number of specimens
+ Tristram's Second Collection, with number of specimens
Bold type in scientific names indicates endemic species that have later been given sub-specific status.

## Table 10: Hawaiian Endemic Birds

| Common Name | Scientific Name (Tristram's time) | Current Scientific Name |
|---|---|---|
| Nene (Hawaiian Goose) | *Branta sandvichensis* | *Branta sandvichensis* |
| Hawaiian Duck | *Anas wyvilliana* | *Anas wyvilliana* |
| Laysan Duck | *Anas laysanensis* | *Anas laysanensis* |
| Laysan Albatross | *Diomedea immutabilis* | *Diomedea immutabilis* |
| Black-footed Albatross | *Diomedea nigripes* | *Diomedea nigripes* |
| Dark-rumped (Hawaiian) Petrel | *Ostrelata phaeopygia* | *Pterodroma sandvichensis* |
| Bonin Petrel | *Ostrelata hypoleuca* | *Pterodroma hypoleuca* |
| Bryan's Shearwater | Pyle 2011* | *Puffinus bryani* |
| Newell's Shearwater | *Puffinus auricularis* | *Puffinus newelli* |
| Hawaiian Hawk | *Buteo solitarius* | *Buteo solitarius* |
| Laysan Rail | *Porzana palmeri* | EXTINCT |
| Hawaiian Rail | *Porzana sandvichensis* | *Porzana sandvichensis* |
| Hawaiian Gallinule | *Gallinula sandvichensis* | *Gallinula chloropus* **sandvichensis** |
| Hawaiian Coot | *Fulica alai* | *Fulica alai* |
| Hawaiian Stilt | *Himantopus sandvichensis* | *Himantopus m.* **sandvichensis** |
| Hawaiian (Black) Noddy | *Anous tenuirostris* | *Anous minutus* **melanogenys** |
| Pueo (Hawaiian Short-eared Owl) | *Asio sandvichensis* | *Asio flammeus* **sandvichensis** |
| Hawaiian Crow (Alala) | *Corvus hawaiiensis* | *Corvus hawaiiensis* |
| Hawai'i Elapaio | *Muscicapa sandvichensis* | *Chasiempis s.***sandvichensis** |
| Kaua'i 'Elapaio | Ridgway 1882* | *Chasiempis s. sclateri* |
| O'ahu 'Elapaio | Stejneger 1887* | *Chasiempis s.ibidis* |
| Mauna Kea 'Elapaio | Pratt 1979* | *Chasiempis s.bryani* |
| Hilo 'Elapaio | Stejneger 1887* | *Chasiempis s.ridgwayi* |
| Laysan Millerbird | Rothschild 1892* | *Acrocephalus f. familiaris* |

| Common Name | Scientific Name (Tristram's time) | Current Scientific Name |
|---|---|---|
| Nihoa Millerbird | Wetmore 1924* | *Acrocephalus f. kingi* |
| Kama'o | Stejneger 1887* | *Myadestes myadestinus* EXTINCT |
| Amaui | Wilson & Evans 1899* | *Myadestes oahensis* EXTINCT |
| Olomo'o | Wilson 1891* | *Myadestes lanaiensis* |
| Oma'o | *Phaeornis obscura* | *Myadestes obscurus* |
| Puaiohi | Rothschild 1893* | *Myadestes palmeri* |
| Po'ouli | Casey and Jacobi 1974* | *Melamprosops phaeosoma* |
| O'ahu 'Alauahio | *Himatione maculata* | *Paroreomyza maculata* |
| Maui 'Alauahio | S. B. Wilson 1890* | *Paroreomyza montana* |
| Kakawahie | *Paroreomyza flammea* | EXTINCT |
| Laysan Finch | *Telespiza cantans* | *Telespiza cantans* |
| Nihoa Finch | *Telespiza ultima* | *Telespiza ultima* |
| Palila | *Loxioides bailleui* | *Loxioides bailleui* |
| Kaua'i Palila | *Loxioides kikuichi* | EXTINCT |
| Lesser Koa-finch | *Rhodacanthis flaviceps* | EXTINCT |
| Greater Koa-finch | *Rhodacanthis palmeri* | EXTINCT |
| Kona Grosbeak | *Chloridops kona* | EXTINCT |
| O'u | *Psittirostra psittacea* | *Psittirostra psittacea* |
| Lana'i Hookbill | *Dysmorodrepanis munroi* | EXTINCT |
| Akikiki | Stejneger 1887* | *Oreomystis bairdi* |
| Anianiau | Stejneger 1887* | *Magumma parva* |
| Hawai'i 'Akepa | *Loxops coccineus* | *Loxops coccineus* |
| Mau'i 'Akepa | Rothschild 1893* | *Loxops ochraceus* |
| O'ahu 'Akepa | Rothschild 1893* | *Loxops wolstenholmei* EXTINCT |
| Akeke'e | S. B. Wilson 1890* | *Loxops caeruleirostris* |
| Ula-'ai-Hawane | *Ciridops anna* | EXTINCT |
| Hawai'i Creeper | S. B. Wilson 1890* | *Manucerthia mana* |
| Kaua'i 'Amakihi | S. B. Wilson 1890* | *Chlorodrepanis kauaiensis* |

| Common Name | Scientific Name (Tristram's time) | Current Scientific Name |
|---|---|---|
| O'ahu 'Amakihi | *Nectarinia flava* | Chlorodrepanis flavus |
| Common 'Amakihi | *Certhia virens* | *Chlorodrepanis virens* |
| Kaua'i Nakupu'u | S. B. Wilson 1889* | *Hemignathus hanapepe* |
| Mau'i Nakupu'u | Rothschild 1893* | *Hemignathus affinis* |
| Akiapolaau | Rothschild 1893* | *Hemignathus wilsoni* |
| Nukupu'u | *Hemignathus lucidus* | EXTINCT |
| Greater 'Amakihi | *Hemignathus saggirostris* | EXTINCT |
| Kaua'i 'Akialoa | *Akialoa stejnegeri* | EXTINCT |
| O'ahu 'Akialoa | *Akialoa ellisiana* | EXTINCT |
| Maui Nui 'Akialoa | *Akialoa lanaiensis* | EXTINCT |
| Lesser 'Akialoa | *Akialoa obscura* | EXTINCT |
| Mau'i Parrotbill | Rothschild 1893* | *Pseudonestor xanthophrys* |
| Apapane | *Certhia sanguinea* | *Himatione sanguinea* |
| Akohekohe | *Himatione dolei* | *Palmeria dolei* |
| I'iwi | *Certhia coccinea* | *Drepanis coccinea* |
| Hawai'i Mamo | *Drepanis pacifica* | EXTINCT |
| Black Mamo | *Drepanis funerea* | EXTINCT |

* Authority of species named later than Tristram's paper

Bold type in scientific names indicates endemic species that have later been given sub-specific status.

— 13 —

# Tristram in Japan

CRANE. (*Grus cinerea.*)

Canon Tristram sailed to Japan in order to visit his daughter Katherine who, at the time, was Principal of Bishop Poole's Memorial Girls' School in Osaka. Katherine had graduated from London University in Mathematics and was the first resident lecturer in Westfield College. Accepted as a missionary in 1888 she remained unmarried and served in the Japan Mission until her retirement from the Church Missionary Society in 1928. There is no record of Tristram providing his daughter with any bird-collecting equipment (Plate 67).

Whilst the account Tristram gave of his time in Japan (Tristram, 1895) loosely followed his journeyings in that country, his recording of natural history, geology and archaeology did not give the detail that had characterised the reports of his expeditions. His reports on the ornithology of Japan were usually in the form of very generalised comments on aspects of the subject rather than commentaries on the species observed. This was probably for two reasons: firstly, that his friend Henry Seebohm (1890) had published a book giving much more detail than anything he would be able to produce in the time he was there and, secondly, his visit was not an expedition of the sort he had previously carried out, but more of a social nature and what we would now describe as a tourist's visit.

Rummaging in a taxidermist's shop in Yokohama, Tristram found the skin of a rare bird— un-named in his account of the discovery—from the Foochoo Islands. The prices asked by the owner were relatively cheap but for this he asked five dollars. Tristram writes of this: "I demurred to the price but I have always found the Japanese are at once fetched by a joke; and so when he told me that the dealer in live birds from across the street asked twenty five dollars for a living bird, I replied, through my daughter, that such a good man as he was worth a thousand dollars when alive, but I would be sorry to give ten for him when dead. The dealer threw himself back, laughing heartily at the joke, and said I might have it for a dollar."

Reflecting on the "bird keeper across the street" Tristram wondered to himself "how the Japanese succeeded in keeping in captivity many species which with us pine and perish in confinement". Great Tits, Marsh Tits, Cole Tits and the Japanese Robin he cited as examples which sang and flourished in their small bamboo cages and maintained a fresh bright plumage in comparison with the same species in Europe, and also China, where they invariably looked bedraggled and unhappy.

The Japanese, thought Tristram, took a delight in their birds and not only their cage birds. They enjoyed having the swallows which abounded in the streets nesting on nearly every house. Chimney Swallows and Red-rumped Swallows, what the Japanese termed 'Bottle Swallows', were equally common, the latter being so named because of the shape of the nest. This is built under an overhanging projection, so that it literally hangs from it and is attached to it, with the entrance as a hole in the side. Tristram describes the area outside the hole being extended as a tube, sometimes a foot in length, which is not usual in the populations of this species breeding in Europe. In contrast, the Chimney Swallow, which is our Barn Swallow, nests on ledges and rafters in Japan, according to Tristram for want of chimneys.

Dining out one evening he was served with pheasant and Tristram asked if this was not the close season for shooting. Having explained that the particular pheasant in question had been trapped, not shot, and having touched upon Tristram's interest in birds, his host produced a second, unplucked pheasant of a different species. This latter was the Green Pheasant, now the national bird of Japan, whereas his dinner was a Copper Pheasant. The Green Pheasant is "symbolic of masculine might and prowess, maternal love and care and also considered a prophet of earthquakes" (Brazil, 1991); despite this, 486,000 were shot in a single year in 1946. Tristram actually experienced a strong earthquake but before seeing the Green Pheasant.

The Copper Pheasant is now uncommon and difficult to find and the shooting of females is illegal. A combination of shooting nearly one million a year, and in Tristram's time, the demand in Paris and England for the feathers for ladies' hats, probably contributed significantly to the decline. As in the West, the Ring-necked Pheasant was introduced from Korea in the Middle Ages as a game bird, which has probably gone some way to preserving the populations of the two native species.

Tristram spent much of his time in Japan travelling with his missionary daughter and on the occasion of their visiting the mausoleum of Iyeyasu, a sometime general and Emperor of Japan, he observed a fresco depicting aspects of falconry above the less interesting cabinets of swords and clocks. The fresco depicted the history of falconry in Japan and Tristram noted the similarity of the hoods, jesses, gloves and other hawking equipment to that used in the West. Few other pastimes show such a similarity and Tristram speculated that the art of falconry had probably spread east and west from Assyria, on the basis of the existence of a sculpture dating from 1700 B.C. and discovered by Sir Henry Layard in Khorsabad, now in Iraq. However, Tristram (1895) quotes a much earlier date from China. Falcons were amongst presents given to the rulers of Japan as early as the Hia Dynasty, which began in 2205 B.C. Harting (1891) asserts that falconry had its origins in the plains of Hindustan, in India, and was introduced by the "Indo-Germanic race before the historic period".

There is a considerable literature in Japan on the art of falconry and Harting quotes fourteen Japanese publications in his *Bibliotheca Accipitraria*. Tristram quotes some of these: "The Japanese writers on falconry mention the goshawk, the peregrine, the sparrow-hawk, the osprey, which they call the pike-catching hawk, the gier-falcon which they obtain from Kamschatka, and last and least, the grey shrike which they have succeeded in training to catch small birds." In the West, the practice of falconry had almost disappeared at the time of Tristram's visit to Japan where there was a similar trend, probably partially due to a society that has come to be more protective of wildlife generally and less inclined to regard it as something to be hunted.

Tristram found the avifauna of Japan to be particularly "interesting to a British naturalist, from its close resemblance to, as well as its marked differences from, our British fauna". The Japanese Pied Wagtail, a much bigger and more striking bird than ours, attracted his attention on a very interesting but not long expedition by path on the riverside to Kamman-ga-fuchi, not far from Nikko.

> Alongside the path the trees and shrubs were ceaselessly visited by little flocks of various kinds of titmice, some identical with, and others very close to, our own. Family parties of the schoolboy's favourite the long-tailed or bottle tit, were seldom absent from view. The representative of the great tit, with exactly the same note as our own, the marsh and cole were everywhere in evidence; and the conspicuous

chestnut, black and white titmouse (*Parus varius*), peculiar to Japan, and its favourite cage-bird, was most abundant of all.

Nearly everything they encountered had a close resemblance to similar species from Europe but nearly always with a difference, less so in the birds.

> That laughing, screaming jay amongst those maples overhead, you would say, was undoubtedly our own jay, to the minutest particular, and yet if you were to handle him he is different, but only by a black streak from his beak to his eye, where our jay is chestnut. And so to the bullfinch, identical at first sight with our bird-fancier's darling and gardener's abomination, voice, flight, nest and eggs indistinguishable; but we shall always find the native of Japan with a ruddy tinge on the back, and less decisive red on the breast, yet bullfinch all the world over.

Another excursion found Tristram's small group on a longer ramble to the Lake of Chusenji, climbing nearly all the way along a fast-running stream and heading for a snow-capped mountain range. Eventually they found themselves in a forest of Cryptomeria, pine fir, maple, alder, oak, birch and larch, all leafless, but the deciduous trees just budding. Here Tristram saw the Japanese Robin and Japanese Hedge Sparrow for the first time "both very like our own, and exactly resembling them on note and habits, though in Japan they are both exclusively mountain birds, said never to be found below 4000 feet (1220 m) and consequently are the rarest of birds in Japanese collections." Arriving eventually at the lake it gave no rise to ornithological comment, nor did the return journey (Tristram, 1895). There then followed several days of sightseeing, which apparently did not involve the sightings of any birds worth comment, but to some extent this was made up for on returning one evening to find the boy who had brought Tristram bird skins on his first evening in Japan. He had been asked to bring more on this second visit and he arrived, together with his employer, with between two and three hundred skins of birds and mammals. Apparently they had been collected for an Englishman, the late Mr H. T. W. Pryer, a lepidopterist and sometime resident in Japan, known to Tristram and from whom he had obtained many specimens.

All the skins which the collector had brought with him were neatly prepared and labelled with the names, places and dates. This surprised Tristram as the Japanese skins he had examined back in England were nearly all labelled 'Yokohama'. All the birds in the collection that was in front of him had been taken in the forest region, on high ground around Nikko, at a height between three thousand and eight thousand feet above sea level. Tristram comments: "No wonder the English writers have gone astray as to the localities of the birds of Japan. It was pretty much as if the dotterels and ring ouzels of Cross Fell should be labelled 'obtained at Liverpool'."

Over one hundred species of birds were represented in the collection offered and for the most part these were in pairs and carefully sexed. As Tristram went through the collection he was delighted by the way the collector and his young assistant attempted, often in sign language, to describe their habits and habitats, and eventually he made an offer, which was accepted, for the whole collection. Apparently the collection had been destined for a dealer in Yokohama who wanted it for an English collector, but Tristram had no hesitation in taking it, as he had previously had a collection in Algeria, which was promised to him, sold to M. Verreaux in similar circumstances. Tristram suspected that few of the birds were shot and that the smaller ones had been limed and the larger snared or taken in traps.

One group of characteristic Japanese birds was conspicuous by its absence from the collection, the cranes. The collector explained that cranes were sacred and that he would not offend either the gods or the Mikado by shooting one. Of the five species "known as belonging to Japan"—there are eight species now on the Japanese list—three are semi-domesticated, the White-headed Crane, the White-naped Crane and the Sacred (Japanese) Crane and are found in parks across the country. Tristram suspected, largely because he saw only one flock of cranes during his stay in Japan, that they would follow the same fate as the British population, that is extinction. Fortunately he was wrong in this; the three semi-domesticated species are locally common and there is now a wintering population of a few Common Cranes. Happily, too, the Common Crane is now breeding again in England.

After arriving in Subashiri, a village close to Mount Fuji, a two hour climb towards the mountain brought them close to the spring snowline. Secluded in a small open space in the middle of the forest, Tristram was able to watch and note the behaviour of some of the rarest birds in Japan:

> It seemed to be a rendezvous of song birds as I sat completely concealed by the foliage of an evergreen shrub, the beautiful Narcissus Flycatcher took its perch on a twig within a yard of my head; the Siberian blue-tail, and, best of all, the lovely Japanese Waxwing, fearlessly hopped about in pursuit of the small butterflies; the Siberian blackbird with its white belly, and the black and white ouzel (*Merula cardis*) perched at the opposite end of the opening, entered as competitors in a singing match, while many a warbler whistled and titmouse chirped unseen. It was an hour's ornithological education such as I have rarely had.

Tristram described the castle in Nagoya, which was the next place they visited, as "The Alnwick Castle" of Nagoya, with many of the rooms being decorated by paintings of a single species of mammal or bird. There was a leopard room, a tiger room and rooms given over to hawks, woodpeckers and pheasants. At their next port-of-call, Gifu, they again found the bird life of importance, where the main summer industry was fishing with cormorants, an activity, like falconry, of great antiquity. This method of catching fish was introduced to Europe by the Dutch at the beginning of the seventeenth century, spread to France and England where it was a favourite amusement of James the First and Charles the First. Tristram had seen cormorant fishing in China, in the Province of Che-Kiang, and found it to be similar to the Japanese method. Young cormorants were taught to feed from the hand and were allowed to fish with a string attached to the foot. They soon learned to return to a call and when sufficiently trained were released with a leather collar attached round the neck to prevent them swallowing any fish they took. When called, they returned to their trainers and disgorged the fish they had taken. Fishing usually occurred at night by using a light to attract fish round a boat, and when the catch was sufficient the neck collars were removed and the cormorants allowed to fish freely. In this manner fishing could be continued throughout the dark hours, the birds being given short intervals of rest from time to time.

Tristram spent most of his time in Japan wandering under the guidance of his daughter and obviously depended a great deal upon her being able to help him converse with the people he met. Whilst it is not actually stated, I get the impression that Tristram would have enjoyed his ramblings in Japan more had he not been restricted by his daughter's missionary zeal, and there can be little doubt that he missed his gun which he used for collecting his specimens. But he did take back to England a good selection of bird skins, probably better documented than any previous Japanese collection, which he could work on alongside his island birds during his old age.

— 14 —

# Twilight of the Great Auk

GOLDEN EAGLE. (*Aquila chrysaëtus.*)

The Sacred Ibis, venerated as the manifestation of the Egyptian god Toth, the god of wisdom and learning, was all but extinct in Egypt when Henry

Tristram went up to Oxford in the autumn of 1840. Common in Pharaonic times, when it was depicted almost always as breeding and feeding in *Papyrus* beds, its demise in Egypt was probably linked to the disappearance of the *Papyrus* swamps. This led to the disappearance of *Papyrus* itself from Egypt, and it was this, rather than the events of two thousand years ago, when thousands of the birds were killed for mummification in the tombs, that led to the effective loss of these birds from Egypt. In fact the birds were domestically bred for the purpose of mummification (Houlihan, 1988) so most of those that were mummified were probably not wild birds.

Whilst during his time at Oxford, Tristram had not yet acquired his Newtonian sobriquet (Sacred Ibis), his later association with the god Toth (also sacred, so another of Newton's puns!), whilst perhaps not being professionally appropriate, was certainly so in relation to his wisdom and learning. Like many other ornithologists of his time Tristram was more interested in a very different bird much nearer to its permanent demise, the Great Auk. Because of its almost widespread and increasing interest to ornithologists, as a result of its rarity, it was surprising that the British Ornithologists' Union did not adopt its name as the title for the new journal in preference to 'Ibis'. This was left to the Americans in 1888, whose journal *The Auk* had its title page depicting the then extinct Great Auk.

At the time Tristram left for Oxford there was a mounted Great Auk (Plate 68) in the museum in Durham, acquired in 1834 or 1835 according to the Keeper, William Proctor (Grieve, 1885). Fuller (1999) questions this date but it is likely to be correct as it falls between Proctor's two visits to Iceland in search of the species which would surely give him a good marker for his memory. Even if it was later, the event must more or less have coincided with Tristram's first visit to the museum and his introduction to William Proctor. By any standards the acquisition must have been appreciated by all in the North-East with an ornithological interest.

Corresponding with Mr R. Champley of Scarborough, himself a collector, Proctor wrote in 1861: "We have the Great Auk in our museum but not the egg. We got the skin from Mr Read of Doncaster, I believe about the year 1834 or 1835. The Rev. T. Gisborne bought the skin in Doncaster for £7 or £8, I believe, but when it was killed or taken I do not know." Fuller records the first known location of this bird in the possession of Fredrick Schultz, of Dresden, who acquired it in that city in 1835 so that going by previous history it is likely to have come from the island of Eldey. It was from Schultz that Read obtained the bird. Proctor later remounted the skin but considered it unsatisfactory and about 1900 Durham University had it cleaned and again remounted. The 'before' and 'after' are illustrated in Fuller's monograph (Fuller, 1999).

Eventually, with the decline of the University museum, the Great Auk moved into the Department of Zoology and took pride of place on the filing cabinet in Professor J. B. Cragg's room. Apocryphally it was then used as a bargaining base for new equipment in the Department, the Professor suggesting the sale of the bird if monies were not to be found. The truth of this story is in some doubt, but on his departure, and the Department having a new head, the sale of the Durham Great Auk was approved by the University authorities in 1977. The temptation to which they succumbed was the possibility of raising the £9,000 which another Great Auk had achieved in the saleroom. Despite vigorous opposition from departmental staff and past students (the present writer amongst them), the bird was sold to a Mr Pilkington of The Dilemma Gift Shop, Knightsbridge for £4,200! It was then loaned by Pilkington to the Glasgow Museums and in 1993 they purchased it for £30,000, clearly demonstrating the rashness of the sale by the Durham University authorities. Tristram would have turned in his grave! The bird is now known as the Glasgow Auk.

When Tristram became interested in the Great Auk it was not quite extinct and there can be little doubt that his new friendship with Newton stimulated his interest in the species. Because of its great rarity, almost all the collectors of the time were keen to acquire specimens which put the species in greater danger. Whilst collectors did not cause the extinction of the species, they completed the demise that fishermen had begun. The Great Auk was the first bird to be declared extinct since the advent of ornithology as a science and the first extinction to be recognised as caused by human activity.

At about the time when Tristram and Newton first met, Drummond had returned to England having made what was probably the last sighting of a living Great Auk on the Newfoundland Banks on his shipboard journey home from Bermuda. At the time, Drummond was not to know this and there were still those who were searching for live birds and, of course, their eggs. Had they found them they would almost certainly have been added to collections on one or other side of the Atlantic.

Within the ornithological community the Great Auk had taken on cult status and, as an icon of extinction, competed with the Dodo and dinosaurs in the world at large. This has been elegantly documented in Errol Fuller's 'Great Auk'—more a celebration of the species than a monograph, the subscribers' edition being a collectors' item, much like its subject. There were earlier chroniclers of the species of which Symington Grieve was the most successful (Grieve, 1885). It drew on observations and discoveries made by other ornithologists. Amongst the foremost of these was Alfred Newton who published extensively on the species in the scientific journals *Ibis* (1861, 1866, 1870, 1898), *Nature* (1885) and *Natural History Review* (1865), and in the *Encyclopaedia Britannica* (1879), his own *Ootheca Wolleyana* (1864–1907) and his *Dictionary of Birds* (1896).

Tristram's interest in the Great Auk clearly started with the Durham specimen in the University Museum but he also acquired an egg (Plate 76) before he became a great friend of Newton in 1853. Newton's manuscript note records it as being collected in 1834 and, having been taken from Iceland to Copenhagen, it remained there in the possession of Nils Kjaerbolling until 1851. It was then purchased by J. De Capel Wise and later acquired by the London taxidermist J. Williams, who sold it to Tristram in 1853 for the sum of £30.

Newton spent a great deal of time researching the history of the Great Auk and he was one of the last ornithologists to accept its demise. Because of his particular interest in the species it was surprising that he did not acquire an egg earlier. On 18 August 1860, Newton wrote to Tristram in great excitement:

> In going to London this very wet day I have picked up the greatest prize an English oologist can meet with. I stumbled upon the scent of it in the subterranean region of Bloomsbury, and after a brilliant

burst in a hansom ran it to ground under the shadow of St Mary-le-Strand, a locality already sacred to the memories of poor old Salmon and his great egg. The long and the short of it is that today I have purchased a Great Auk's egg, one whose existence was previously unknown to me. I felt bound to rescue the Andromeda from being chained in the sunshine of Gardener's window, but I must confess she is not remarkable for her good looks, though I have seen worse, and I am glad to say her antecedents are likely to prove extremely interesting. (Wollaston, 1921)

This was certainly the case, and in more ways than one!

Both Woolaston (1921) and Fuller (1995, 1999) have given very full accounts of the history of the egg. Firstly, and this is not in doubt, it was produced by the same bird that laid the egg (Fuller No. 60) in the possession of the Museum of Comparative Zoology of the University of Harvard. Apart from being marked in a very similar way, both eggs possess a semi-spiral depression at the pointed end. According to Newton this depression was caused by the musculature in the duct through which the egg passed, in a plastic state, before it was laid. Secondly, it appeared that at one time the egg had been in the possession of Mr J. D. Salmon, a well-known egg collector, who bequeathed his collection to the Linnean Society. When the Society eventually took possession of the collection, the Great Auk's egg had been replaced by a poorly painted swan's egg! Furthermore, the premises from which Newton purchased the egg (from a Mr Calvert) had previously been those occupied by Salmon, so the possibility arose of some deceitful dealing by Calvert.

Quite clearly the egg may well have been sold by Salmon before he died, or by his executor after the event, so there was no suggestion that Newton had behaved in any way improperly, and he retained the egg which is now in the University Museum of Zoology in Cambridge (No. 6 of Fuller's listing). The Catalogue of the collection which had passed into the possession of the Linnean Society had had the page relating to the Great Auk's egg removed before it came into the Society's possession. This was first noted by Mr Bidwell who purchased the Catalogue from Mr Calvert. The former transferred the Catalogue to the Linnean Society in 1891.

Something over a year later, Newton made an even greater discovery. Visiting Surgeons' Hall to examine Richard Owen's dissection of a Great Bustard, he passed a case of eggs which contained ten eggs of the Great Auk. Newton in writing to his brother Edward on Christmas Day 1861 (Letter in Wollaston, 1921) stated: "Of course I hardly obtained credence from my friends but the next day I took Tristram and Sclater and Simpson, and we all four had the case opened and handled the eggs which are neatly sealing-waxed to the boards." Wollaston records that Newton did not obtain any of the ten eggs, but his collection, now in the University of Cambridge, eventually contained seven eggs of the Great Auk.

Again on this occasion Tristram was the first of Newton's friends to be told of Newton's triumph, but apparently he was satisfied with his single rather undistinguished egg, but he too retained a long-time interest in the Great Auk. As has already been mentioned, the first stop Newton made on his return from Iceland after investigating past nesting sites of the Great Auk, was in Castle Eden to visit his friend Tristram immediately after disembarking. Tristram was thus the first to hear of what Newton had found, or not found, in his examination of possible nesting sites in Iceland.

Both Newton and Wolley who accompanied him, had high hopes of success on the expedition to Iceland, but they were to be disappointed, both in terms of finding living birds and even in getting to their former nesting sites. Their journey to Iceland by boat was uneventful and after leaving Thorshaven in the Faroe Islands about midnight on the 24 April 1859 they passed the Westermann Islands in the early morning of the 26 April. About noon the Meal Lack, as Newton first referred to the Island of Eldey, was sighted some two miles away. "On the landward side runs a low shelf of rock, whereon the Great Auk is supposed to have bred. Outside, at about four times the distance, lies a small, low skerry which had a very inviting appearance, but the water is said at times to go right over it . . . We looked, of course, for the Geier Fugle, but in vain." (Letter dated 2.5.1858 in Wollaston, 1921)

Off the Iceland Cape Reyjanes and running south-west as a group of four main islands, of which Eldey is closest inshore, there comes Eldeyjardrangr, then Geirfuglasker (Garefowl Island) and Geistfuglasker. Bad weather kept Newton and Wolley in Reykjavik until the 19 May when they moved to Cape Kirkjuvogr, near to Cape Reyjanes, from where they hoped to access Eldey. Here Newton wrote again to his brother: "Wolley is much more sanguine about success than I am, and I think more than he has a right to be; but at the same time I am not more desponding than I have always been about it." (Letter in Wollaston, 1921)

Their landlord in Kirkjuvogr had taken two birds and two eggs (one broken) on Eldey in 1844, and on two previous visits seven and twenty-four birds, but on a visit in 1856 had found nothing. In fact, there was never much chance of finding either birds or eggs in 1858 but they were not to be given the chance. The weather did not improve, and a crossing to Eldey was impossible. Back home on the 16 August Newton wrote: "The result then, in short was nothing. Not one day of the whole two months we were in Kirkjuvogr was the sea sufficiently calm to have allowed us to land, even had we gone out, and we have come back knowing no more than when we started whether the Great Auk is living or dead." (Letter in Wollaston, 1921)

A fact which Newton brooded on, and which he must have discussed with Tristram, concerned the collection of the last two birds and two eggs (one broken) on Eldey. Auks normally breed in pairs so that two eggs require four birds to produce them. Where then were the two other birds that formed the two pairs when the collection was made? Were they at sea? If so, this account was not of the last two birds associated with Eldey. A further point which exercised their minds was the source of all the Great Auks' eggs that are in collections. Certainly on Eldey Newton could account for only as many as six eggs being collected over a period of thirty years. Dealers gave large sums for bird skins but "never cared much for eggs". The eggs must have come from another island. Fuller records seventy-seven eggs in collections, only two of which definitely came from Eldey, and probably five more were from there, a very similar estimate to that of Newton.

By the time that Newton had begun to collect "any and all information" on the Garefowl, as he termed it, he did so with the objective of producing an authoritative monograph. He limited the breeding area of the Great Auk in the north Atlantic south of the Arctic Circle and he documented the known skins and eggs. Historical and archaeological evidence was collected by him and he showed it to have been a breeding bird on both sides of the Atlantic. On the western side Newfoundland, particularly Funk Island, in the north the islands off Iceland, the Faroes on the eastern side and in Scotland St Kilda and possibly the Orkneys (Papa Westray) were all possible breeding sites. Further south on the eastern side of the Atlantic archaeological remains demonstrated their presence, at least outside the breeding season, though there is some evidence to suggest that up to 1790, Great Auks occurred in summer in the Isle of Man. Similarly on the eastern side of the British Isles there are records of archaeological remains discovered in 1878 at Whitburn Lizard, Northumberland and of a live bird on the Farne Islands, which was caught and kept in captivity for some time (Hancock, 1874).

Newton did not publish his promised monograph on the Great Auk largely because in 1885 Symington Grieve published his, 'The Great Auk or Garefowl *Alca impennis,* Linn., its history, archaeology and remains'. Whilst this is a useful contribution to the subject it is not the monograph that Newton would have produced. Rather embarrassingly Newton was asked to review it and as usual he wrote what he thought. It was not a good review but Grieve accepted it and the two apparently remained on reasonably good terms.

Such a review could not be addressed to the latest monograph on the Great Auk (Fuller, 1999) which provides images of all the known mounted birds which still exist and many of the seventy-five eggs which still remain in collections throughout the world, and two eggs which have been destroyed. This beautifully illustrated monograph concludes with a Turneresque view entitled 'The evening of June 2nd 1844' on Eldey, and illustrates an important question: what happened to the last two Great Auks and the egg(s) which were collected? Previous writers have found no traces of what happened to these specimens. One egg was broken and the remaining egg, and possibly the broken one, were then collected. The two birds and the egg(s) were taken by Sigurdr Islefsson, Jon Brandson and

Ketil Ketilson. The party was led by Vilhjalmur Hakanarsson, Newton and Wolley's host at Kyrkjuvogr. The two Britishers were able to speak to twelve of the fourteen men in the collecting party, the other two being dead. It was from them that Newton gathered his information about the capture of the birds. One witness held that there were two eggs.

The two birds (and the egg(s)) were sold for the equivalent of nine pounds to Christen Hansen and they then passed to Herr Moller, an apothecary of Reykjavik, who had the birds skinned. Newton does not mention Moller's part in this but records that "the (skinned) bodies are now preserved in spirit in the University of Copenhagen, but respecting the ultimate fate of the skins I am not quite sure." Fuller records that Moller had a portrait of one of the birds painted by a French artist named Vivien which was seen by Newton and Wolley. Gaskell (2000) goes further, recording that Etatsraad (Councillor) Eschricht took the skinned bodies from the Zootomical Museum to the Zoological Museum in Copenhagen University where Japetus Steenstrup identified them as male and female, a fact which was later confirmed by dissection in 1940 (Fuller, 1999). It would seem likely then that the two birds killed would be the owners of one of the eggs.

Meanwhile, in Flensberg, on the Danish border of Germany, the apothecary and dealer Mechlenberg had reportedly two birds and an egg for sale and these were offered to John Hancock for a price of £24 (Bolam, G. 1912, Letter from Hancock). These were said to be a pair with their egg brought from Iceland "a year or two previously". Even at that price it seems unlikely that they had not sold previously, so they were probably recently taken. Hancock offered £10 for one bird in summer plumage and the egg and obtained them for this price. Newton (1864–1907) reports that the egg "obtained by the intervention of Mr Sewell in 1844 or 1845 . . . is more than likely one of the last to be taken on Eldey. It is in a fine condition and was figured by Mr Hewitson . . . in 1846, before it had time to lose its freshness", again an indication that it had been recently laid.

Since nearly all the Great Auks and their eggs were collected in Iceland for sale abroad and not in a very systematic manner, there are very few records of eggs taken being attributable to any particular bird. It is therefore of interest when such an event occurred and it would be particularly attractive to potential buyers, who in the main were collectors. The coincidence of two groupings of two birds and an egg being on the market at the same time is unlikely.

The fact that the eggs were recently laid and linked with two adult birds, associated with the fact that both groups were from Iceland, were both associated with apothecaries dealing in eggs and skins, together with the known fact that the specimens collected on Eldey in June of 1844 were probably a pair (at least a male with another female) suggest that here we are dealing with a single grouping of two birds and an egg. With only seventy-eight known birds and seventy-five eggs, what are the odds of two groupings, each of a pair with an egg, being on the market at the same time, particularly when the original collectors so seldom knew which skin was associated with a particular bird? As all the literature shows there is little evidence of any data being associated with most of the specimens collected. There is at least an argument that Hancock's egg (No. 26 of Fuller) and Hancock's Auk (No. 10 of Fuller) are from this same grouping of birds and egg taken by the fisherman on Eldey in 1844.

Whilst all this was taking place, Tristram was in Bermuda but he probably heard of Hancock's acquisition and went to examine it on his return to the North-East in 1849. So too must Newton have heard about Hancock's egg and he probably examined both the bird and the egg. Those addicted to the Great Auk story—and most ornithologists at the time were—kept abreast of what was happening to the species. This was the first time that a bird on the British List had become extinct within living memory. It certainly brought home the fact that species do become extinct and it also attracted attention to the converse; do new species come into being and if so how? Undoubtedly these questions became major topics of conversation between Tristram and his ornithological associates, particularly with Newton. For the first time it became obvious that humans had brought about the final extinction of another organism. It is arguable that the Great Auk population was on the decline before, as Evans (1900) put it, "the species (was) extirpated chiefly by the persecution of fishermen, but subsequently by collectors." The birds were collected in their hundreds and salted down aboard ship by British and Portuguese fishermen. Annual massacres were continued over a number of years until such a time as there were few birds left to kill, and until the collectors took charge and the last Auk was gone. Probably the last few birds survived a little longer than the Eldey birds but in the British Isles the last dates of capture were recorded in Orkney in 1820, St Kilda in 1821 and in Ireland in 1834 (Evans, 1900). Since then, and helped by the Victorian desire to possess trophies, the Great Auk has gained iconic status and its fate probably prompted both Tristram and Newton to form what might be termed the grassroots of the conservation movement. Perhaps the Great Auk did not die in vain.

Plate 67: Henry Baker Tristram about the time of his visit to Japan. He grew his beard "for health reasons" and had to obtain a special dispensation from his Bishop to do so. He found it useful to keep his throat warm during his expeditions but kept it for the rest of his life.

Plate 68: Great Auk *Pinguinus impennis*. The once Durham Great Auk is now in the possession of Glasgow Museum. Reproduced with the permission of the Glasgow Museum authorities. © CSG CIC Glasgow Museums Collection.

Plate 69: Norfolk Island Parakeet (Kaka) *Nestor meridionalis productus*. Once named the Long-billed Parakeet, the extension of the upper mandible is now recognised as a malformation. The skin is in the Liverpool World Museum, originally from the collection of Lord Derby. Reproduced from *The Bulletin of the Liverpool Museums*, 1898, Vol. 1: frontispiece, with the permission of the Liverpool World Museum authorities.

Plate 70: Labrador Duck *Captorhynchus labradorius*. A specimen of this extinct duck was given to Tristram by his long-time friend from Bermuda, by then Lt. Col. Wedderburn. It was not "stolen from his greaving widow" as claimed many years later (see text). Reproduced from Rowley's *Ornithological Miscellany* (1875–7).

**Plate 71:** Large Ground Finch *Geospiza magnirostris*. This bird in non-breeding plumage has a massive bill but the specimens first given this name had even larger bills (see text).

**Plate 72:** Warbler Finch *Certhidia olivacea*, the smallest of Darwin's Finches. Comparing this with the Large Ground Finch it is not surprising that Darwin did not see a relationship.

Plate 73: Duchess Lorikeet *Charmosyna margarethae*. Named by Tristram "in compliment to the bride of H.R.H. the Duke of Connaught". Reproduced with the permission of the British Ornithologists' Union, www.bou.org.uk from *Ibis* 1879, Vol. 21, Plate 12.

Plate 74: Bare-legged Scops Owl *Otus insularis*. Named by Tristram in 1880, it "is particularly interesting as being the first living species of Owl discovered in any of the Mascarene Islands." Reproduced with the permission of the British Ornithologists' Union, www.bou.org.uk from *Ibis* 1880, Vol. 22, Plate 14.

Plate 75: White-collared (Mangrove) Kingfisher *Todiramphus chloris tristrami*. Originally given specific status (Layard, 1880 in *Ibis* p. 459) as *Halcyon tristrami*, it is now considered to be one of 49 sub-species recognised by del Hoyo et al., (1992–2013) in this morphologically very variable species. Reproduced with the permission of the British Ornithologists' Union, www.bou.org.uk from *Ibis* 1880, Vol. 22, Plate 15.

— 15 —

# "What's Hit is History, What's Missed is Mystery": The Collections of Skins

SAND-GROUSE. (*Pterocles senegalensis.*)

In his *Catalogue of a Collection of Birds*, (1889) Tristram states that his collection was begun in 1844, with specimens collected in Switzerland and northern Italy. It is strange that his early collections, made as a schoolboy and as an undergraduate in Oxford, gathered under the guidance of William Proctor, the sub-curator of Durham Museum, did not form the basis of this collection, but had disappeared by the time the collection was begun. Such early collections are often the prize possessions of their owners and it would be surprising if this were not so in Tristram's case. Nevertheless his collections had disappeared and he was to begin again and not for the last time.

Like most ornithologists of his time, Tristram began his collection with birds from his home area and from the British Isles in general, followed by those acquired during the three years he spent in Bermuda. On his return journey to Northumberland he stopped off in North America, in Canada and then for a short time in northern USA. There followed his sojourns to the Mediterranean, North Africa, the Great Sahara, Tunisia, Palestine and Syria, when, over a period of forty-five years, he assembled one of the greatest of all natural history collections ever made by a private individual. He described this as "the limited leisure of forty-five years", but one might wonder how he defined the term "limited"! There can be no doubt that he put considerable effort into the acquisition of his skin collection. Identifying what he had collected was a very time-consuming measure. There were no field guides and other literature was not as accessible as it is now. The status of previously unrecorded birds then took much longer to determine. Having done so, it often happened that another worker had got into print first, so invalidating any name attributed to the species in question after that first published. This would invalidate any name attached to the specimen following the rule of (chronological) priority.

Skinning the bird, often in the field, preserving it for the cabinet, sorting, labelling and cataloguing specimens also took up a great deal of time. There were also the eggs that he collected to deal with, blowing and packing them safely and all the while making other collections of plants, molluscs, fish, amphibians, reptiles and mammals. All this was a phenomenal undertaking and Tristram's "leisure hours" must have extended long into the nights, particularly when he was involved with one of his numerous expeditions.

There are various estimates of the size of Tristram's skin collections. During the course of preparing the Catalogue of the collections, Tristram (1889) wrote that it consisted of over 17,000 specimens, but it is known that not all of his first collection went to Liverpool (accession date, 2 April 1896) as there are specimens prepared before this date which went with the second collection to Philadelphia. Baker (1996) quotes a figure of 16,240 specimens as being in the Liverpool Museum at the date of his writing "based on a recent catalogue". Apparently at the time of the accession to the Liverpool Museum, no count was made. There are addenda to the catalogues of which Tristram writes: "The names of species in the Addenda, not included in the Catalogue, are printed in *thick* type. The names of species already in the Catalogue are printed in *thin* type." All the specimens in the Addenda are in addition to those included in the main Catalogue. It is unlikely that the specimens transferred from Durham University to Liverpool Museum in the early 1960s are included in the Catalogue, but short of comparing all the individual skins with the listings there is no means of checking this. Again, after an expedition to Palestine in 1864, Tristram gave some 265 specimens that had been collected there to the Natural History Museum, then in South Kensington, but this was by no means the total number collected (Baker, 1996). Tristram also gave skins from the Pacific Islands to the Museum some years later.

Specimens which Tristram collected found their way into numerous collections world-wide as he exchanged them with many individuals and museums. Bodenheimer (1956a) comments that Tristram's Catalogue lists "birds given or exchanged by almost every known explorer of his time" and seventeen of the more important. Baker (1996) draws attention to the fact that Liverpool Museum produced a compilation of donors and date index in 1978 which showed that Tristram received material from a wide range of sources, comprising 781 donors and collectors. Tristram's contacts ranged throughout the world and there were few prominent ornithologists with whom he could not claim an acquaintance. He contributed a short biography to the series of leaflets entitled *Well-known Collections and their Collectors*, written sometime after 1891. He wrote: "My collection of birds went to Liverpool in 1896 containing about 20,000 specimens." His next sentence, "I never counted my collection of eggs but it was over 20,000; I should say 24,000 at least", implies that he had counted the skins.

Despite claiming that his collection went to Liverpool in 1896, records remain of material still arriving in Liverpool in 1900, so that Tristram's estimate of 20,000 is quite likely to be more accurate. Bodenheimer (1956a) found difficulty in tracing Tristram's collections during two summers he spent in Durham. The lack of a complete run in the university library of the *Ibis*, in which Tristram published most of his papers, and in which the location of his main bird collection was indexed (*Ibis* 1897, 488) obviously made this difficult for him. Tristram's collection going to Liverpool was also mentioned in his Obituary in the *Ibis* (1906, 602), in the Proceedings of the Royal Society B, 8:xlii–xliv, and in the *Geographical Journal* (1906, 27:511–12).

The second collection of birds which Tristram began in 1896 contained "2717 species and nearly double that number of specimens" (ca. 5,400), at the time of his completion of the text for his *Well-known Collectors* leaflet, so that the date of publication of the leaflet was probably 1904 or 1905. Liverpool Museum purchased the first collection for £3,150.

For the second collection of skins made between 1889 and Tristram's death in 1906, there is clearer information available. This collection was purchased, after Tristram's death, by the Academy of Science in Philadelphia and now consists of 5,912 specimens. However, the figure quoted by Baker (1996) may be more accurate, as some years ago, when the data were digitised, many people were involved and many mistakes made. Apparently there may well be 7,000 Tristram specimens in the collection in Philadelphia (Nate. Rice, personal communication).

Another small collection of Tristram skins is present in the National Museums of Scotland, in Edinburgh (R. McGowan, personal communication); 26 of 77 specimens in a small collection presented by Harry Holland are attributable, through the family, to Tristram, the others to Jardine. None appear in Tristram's Catalogue (1889), though none of those dated were collected later than this; probably they were passed on to another member of the family or were simply separated from the main collection as probably happened with those found by the present writer in Durham, now in the main collection in Liverpool.

When such collections as those of Tristram are accessioned into museum collections, they are no longer kept in their original cabinets. This would be impracticable from the point of view of the amount of space they occupy. Now, more modern cabinets, designed to conserve space, are utilised and specimens brought together largely on a basis of species. In large collections, such as that in Liverpool and Philadelphia, Tristram's birds are distributed throughout the museum collections but ideally are accessed through computer listings. Theoretically the collections can be accessed easily under any heading, such as collector, place of collection, date, species, etc. Obviously the data are derived from the labels on the individual specimens so that no labels should ever be removed and new labels, if they are necessary, should be added and old ones not replaced. Not all museums have the facilities or the manpower to conserve their collections ideally, so that mis-labelling, entry of data incorrectly to the computer listings and mis-filing in cabinets can create problems for researchers. It is therefore not surprising that an assessment of the total numbers of skins or eggs accumulated by a past collector cannot be determined accurately. However, it seems likely that Tristram accumulated over 20,000 skins, and he claims in his 'Collections' pamphlet that more than this number went into the museum in Liverpool.

Whilst Tristram collected many of his specimens himself, others were exchanged and, according to him, some purchased. In the preface to his Catalogue (1889) he writes: "On nothing do I look with more satisfaction than the names of the naturalists, explorers and travellers whose labours have aided in enriching the collection." Of the twenty members of the British Ornithologists' Union listed in the first volume of the *Ibis*, T. C. Eyton, F. duCane Godman, J. H. Gurney, Alfred Newton, Sir Edward Newton, Lord Lilford, Osbert Salvin, P. L. Sclater and John Wolley, nine in all, were credited with adding to the collection, as were many other famous British ornithologists, many of whom were later elected to the Union, amongst them: J. Biddulph, W. T. Blandford, Sir Walter L. Buller, E. Coues, C. Danford, Charles Darwin, H. E. Dresser, H. T. Elwes, H. W. Fielden, P. S. Godman, John Gould, J. H. Gurney Jr, John Hancock, J. E. Harting, V. Legge, Prof. T. Salvadori, G. E. Shelley, Howard Saunders, Henry Seebohm, Prof. H. Schlegel, Sir A. C. Smith, and R. G. Wardlaw-Ramsey. From abroad came many contributions from other famous ornithologists, for example Dr A. E. Brehm, Pere David, E. A. Eversmann, H. Gatke, Dr A. Habel, Prof. K.

J. Ghartlaub, T. Van Heglin, T. C. Jerdon, T. Kleinschmidt, E. L. Layard, V. P. E. S. Ruppel, J. P. Verreaux and A. R. Wallace.

Several of Tristram's British contacts worked abroad, amongst them Sir Edward Newton (Mauritius and Jamaica), E. L. Layard (Fiji and New Caledonia), Robert Swinhoe (China coast) and Dr John Kirk (Zanzibar), and they too provided specimens for Tristram's collections.

The vast numbers of other collectors with whom Tristram dealt can be found in the Catalogue (1889), credited opposite the specimens they collected and in the listings in the Philadelphia Museum. In most cases these names also appear on the labels in the various skin collections. Many of these collectors were not ornithologists or even people interested in birds, that is before Tristram engaged them. His being a Governor of the Church Missionary Society brought him in contact with missionaries who were to be sent far and wide. Tristram, not being one to miss the chance of acquiring a new bird, equipped each of his missionaries with a gun, ammunition, skinning instruments, preservatives and labels, and sent them forth to spread the word. Clearly, high on their list of duties would be the enlistment of a bird catcher! He also enlisted support from the Royal Navy, particularly through Lieutenants Gunn, Piers and Richards. Lieutenant Henry Piers was Tristram's cousin and he was commissioned (by Tristram, not the Navy) to enrol more maritime collectors. Piers was surgeon on the *Investigator* during the unsuccessful search for the lost Franklin Expedition, and found time to collect what was possibly only the second egg to be found of the Labrador Duck. This is now in the collection at Tring as are other rarities which Piers brought back from the three years' search. These included the Arctic Black Guillemot *Cepphus grylle mandtii* (three specimens)—now regarded as a sub-species of the Black Guillemot – and Ivory Gull (two specimens), both from Melville Island, the former collected in the summer of 1852 and the latter in 1853.

All this clearly had to be financed: the bird equipment for the missionaries, postage from abroad, other correspondence, and straight purchases, for example from Mr Kanaga in Japan, where Tristram gave "a reasonable price for the whole collection" offered (one third of what Kanaga would have charged a Yokohama customer!). Tristram was known to have paid as much as £10 for a single skin of a Hawaiian Hawk (Mearns and Mearns, 1998). It therefore comes as a surprise that his grand-daughter Eleanor claimed that "he only spent £100 on his collection"! Eleanor's account of her grandparents was written when she was eighty-three years of age, and her memory may not have been what it was. She did make several errors in her 'Recollections' and some of her statements are contradicted in Tristram's written accounts. However, in relation to the comments about his collection, had the matter cropped up, he would, probably have subtly changed the line of the conversation! Even so, Tristram was unusual amongst the great collectors in that he had no substantial purchasing power, whilst others like Salvin, Sclater and Godman had extensive private means. His success in amassing over 24,000 skins is a reflection of his own extensive and single-minded efforts and, more notably, in that his collections extended to plants, insects, molluscs, and geological and archaeological specimens. He was a truly remarkable all-round naturalist, but a great ornithologist.

Tristram was very proud of his collections and some of the birds he obtained are worth special mention. In his introduction to the Catalogue (1889) he makes special reference to several species thought to be extinct, notably the Norfolk Island Parakeet *Nestor productus,* the Labrador Duck *Camptolaemus* (now *Camptorhynchus*) *labradoricus* and the Rarotonga Monarch Flycatcher *Monarcha* (now *Pomarea*) *dimidiata.*

Tristram's specimen of the Norfolk Island Parakeet (now Kaka) (Del Hoyo et al., 1997) or Norfolk Island Parrot (Greenaway, 1958) is now considered to be either a member of a super-species with the Common Kaka (the former authority) or conspecific (the latter authority). Tristram in his Catalogue enters the specimen as *Nestor productus*, a single specimen taken on Philip Island in 1788 which passed into the hands of Dr W. E. Leach who was trained in medicine but was later employed in the British Museum and after whom the Leach's Petrel was named. Subsequently purchased by Lord Derby the parakeet was "received from the Trustees of the Derebianum Museum in Liverpool, in exchange" by Tristram in 1873. Later the specimen returned to Liverpool with the rest of the Tristram collection between 1889 and 1898, at which latter date it was included in a *Catalogue of the Parrots* (Forbes and Robinson, 1898) in the first volume of the Bulletin of the Liverpool Museum (Plate 69). Tristram (1892) was surprised to find that his specimen of *N. productus* more closely resembled the description of *N. norfolcensis* von Pelzen given by Salvadori (1891) in the British Museum Catalogue of Birds, and referred to by Latham (1822) as the Long-billed Parakeet. *N. productus* was thought to be found only on Philip Island, which is off Norfolk Island to which *N. norfolcensis* was supposedly limited. Greenway (1958) regards the two as conspecific and the corneous extension of the upper mandible as an aberration. This explanation having now been adopted, the specific name must now revert to *productus* if the fifteen specimens surviving (one in Australia, three in the USA and eleven in Europe) are considered to be specifically distinct from the Common Kaka, and *N. meridionalis productus* (Gould, 1876) if it is not. Whilst Greenaway favoured the latter course, more recent authorities Foreshaw and Cooper (1975), del Hoyo et al., (1997) and Higgins (1999) favoured the former. Whether a species or a sub-species it became extinct in the wild sometime before the last specimen died in captivity in London, shortly after 1851. Thus Tristram's specimen was not quite as rare as he thought.

When Tristram was in Bermuda in the late 1840s he struck up a friendship with Lieutenant J. W. Wedderburn and over the years they corresponded. Some forty years after they first met the then Lt. Col. J. W. Wedderburn very generously presented Tristram with the skin of an immature Labrador Duck which both Tristram (1889) and Newton (in Rowley, 1875-7) claimed as the last specimen to be collected before its extinction. Wedderburn had shot this bird in the autumn of 1862, off Halifax, Nova Scotia. However, it was not the last of its kind as Phillips (1922) records several alive and shot in the USA after this date:

- 1866 Long Island, Great South Bay. Letter from D. G. Elliot in Rowley, 1877.
- 1871 Grand Manan Island. Shot by S. F. Cheney. Now in the Smithsonian Collection.
- 1875 Long Island. Shot by J. G. Bell. Now in U.S. National Museum, No.77126.

- 1878 Elmira, New York. Shot by a schoolboy, 12 Dec. Recorded by Dr W. H. Grey.

D. G. Elliot recorded several alive between 1852 and 1862. Phillips (1922) records fifty specimens in museum collections, twenty-eight in the USA and twenty-two in Europe.

Referring to a letter from Tristram to W. E. Brooks dated "Dec. 20th 1885" (NEWHM:2002:H1047.9), the grandson of W. E. Brooks, Allan C. Brooks, claimed quite erroneously and probably quite libellously, that when paying a condolence call on the wife of the first owner, Lt. Col. Wedderburn who had shot the duck, Tristram "literally stole the bird from the greaving [sic] widow". Nothing could be further from the truth. There was no grieving widow, nor was Wedderburn dead as Gladstone (1927) records in a letter to *Ibis*. Wedderburn was very much alive when he gave the duck to Tristram as a letter from Mrs Wedderburn to Gladstone, dated "July 21st 1913", makes clear, though she clearly disapproved of her husband giving the bird to Tristram.

Little is known about the biology of the Labrador Duck (Plate 70) which is thought to have become extinct by the late 1870s. The fullest accounts occur in Rowley (1877) and Phillips (1922), the latter, peculiarly, concentrating on what is not known about this duck, now thought to be more closely related to the scoters than to the eiders. According to Rowley, in the 1830s and 1840s it was to be found not uncommonly in the markets of the Atlantic coast of North America: "In the New York markets there would at times be a dozen of them, and then for a few years not one." Continuing his account: "Not many years ago it was a common bird along our coasts." The species was clearly well-known to gunners who had their own names for the species: 'Skunk Duck' along the Labrador coast, 'Pied Duck' in Carolina and 'Sand-shoal Duck' in New Jersey. Audubon (1843) was familiar with the species and commented: "It is usually seen in flocks of from seven to ten, probably the members of one family." He further comments: "A bird-stuffer whom I knew at Camden had many fine specimens all of which he had procured by baiting fish hooks with the common mussel on a 'trot-line' sunk a few feet beneath the surface." Once on the hooks the birds drowned.

According to Audubon's account, the ducks occurred as far south as Chesapeake Bay in winter and as far north as Labrador in summer, but it was never common. It could well have been a species on its way to extinction before the depredations of humans in the form of shooting, trapping and egg collecting. Phillips (1922) favours the loss of food supply as the most likely explanation, coupled with the bird being a specialist feeder with a bill very much adapted to specialist feeding on molluscs and crustaceans. All these are speculative suggestions to which might be added another, genetic alteration, according to Mayr (1951) and Fisher (1953). These authors quoted examples of the Serin Finch in Europe and the Collared Dove expansion from the Middle East resulting from a mutation, and clearly a similar mutation could lead to extinction through either reduced fecundity or increased mortality or both.

Tristram's third "extinct" species, of which he had in his collection two males and a female, was the Rarotonga Monarch-Flycatcher *Monarcha dimidiata,* now *Pomarea dimidiata* and between times *Chasiempis dimidiata.* This species was recorded as extant in the *British Museum Catalogue* (Sharpe, 1879) and in the *Handlist of Genera and Species of Birds* (Sharpe, 1901), though in both Tristram's Catalogue (1889) and its review in *Ibis* (Sclater, 1890) it was recorded as extinct; del Hoyo et al., (2006) recorded it as "Thought to be extinct" in the early 1900s, but in a 1983 survey twenty-one individuals and two nests were recorded. In 2006 an estimated population of 280 individuals had survived the severe cyclone of 2005. Despite not being extinct the small population of this species is regarded as endangered and a conservation programme supported by local landowners is in place.

Apart from the extinct and supposedly-extinct species which Tristram regarded as his prize specimens, he obtained, apparently directly from Dr A. Habel, some thirteen specimens of what we now refer to as Darwin's Finches. Habel had collected these in 1868, he being the first to follow Darwin to collect birds on the Galapagos Islands.

Tristram was fortunate to obtain five of the fourteen species of finch to be found on the Islands:

- Small Ground Finch *Geospiza fuliginosa*, Male and Female, Indefatigable, 16.10.68
- *G. parvula*, male, Abingdon, 30.12.68? Bindloe, 30.12.68
- Medium Ground Finch *G. fortis*, Female, Abingdon 14.12.68. Female, Bindloe 3.11.68
- Large Ground Finch *G. strenua*, 2 Males, Abingdon. n.d. Female, Bindloe, 16.11.68
- Cactus Finch *Cactornis scandens*, 2 Males, 1 female, n.d. Indefatigable, 16.11.68
- Large Tree Finch *Camarhynchus variegates* (now *psittaculus*) 1 Male, Abingdon, n.d.

*Geospiza strenua* is no longer recognised. This form is now named *G. magnirostris* and the form which originally bore this name was a larger-billed bird no longer present in the islands and now extinct. This may have been a different species but is more likely to have been a very large form of the present *G. magnirostris* and conspecific with *G. strenua* which had the bill size of present day *G. magnirostris* (Plate 71).

After these had gone to Liverpool Museum he acquired a further eight specimens of five species for his second collection:

- Small Ground Finch *Geospiza fuliginosa*, Male, Abingdon, 1897. Coll. Harris
- Large Ground Finch, *G. magnirostris*, Female (Hall), Male (Beck), Abingdon, 1897.
- Small Tree Finch *Camarhynchus parvulus*, Male, Charles, 1897, Coll. Hall
- Vegetarian Finch *Platyspiza crassirostris*, Male, Chatham, 1881 Female, Duncan, 1892, Coll. Beck
- Warbler Finch *Certhia olivacea*, Male, James, 1897, Coll. Harris

Charles Harris and Rollo Beck were well-known collectors and visited the Galapagos together in 1897, the former as leader of the expedition, the latter as assistant collector.

In all, Tristram at one time or another possessed the skins of seven of the fourteen species of Darwin's Finches recognised at the time. Later the two species of Warbler Finch (*Certhia olivacea* and *C. fusca*) (Plate 72) were combined as a single species but in 2009 were again split on the basis of DNA research. At this time fourteen species of Darwin's Finches were recognised in the Galapagos, but subsequently the Mangrove Finch was considered to be extinct, which would have left only thirteen extant species on the islands. However, a small population of the Mangrove Finch has been found and conservation methods are now in place in an attempt to preserve the species. There is a fifteenth species, the Cocos Finch *Pinaroloxias inornata* which is unique to the Cocos Islands, 600 miles to the north-east of the Galapagos.

The importance of Darwin's Finches (they were only given the name in 1936 by P. R. Lowe) was not realised until they had been examined by John Gould, after the return of Darwin from the Galapagos. Gould found them to be a group of fourteen related birds which he placed in the Sub-Family Geospizinae, and with the exception of the Cocos Finch are restricted to the Galapagos Islands. Whilst no mention of these birds is made in the Darwin/Russell paper of 1858 or in Darwin's *Origin of Species* (1859) they were to become central to the evolutionary theory and to the species concept. This made them effectively the most wanted specimens for either public or private skin collections. The skin collectors Beck (2) and Harris (1) both supplied skins of Darwin's Finches to Tristram, and it is surprising that he did not acquire more for his second collection in the light of the numbers which Lord Rothschild received. From the expedition to the Galapagos led by Harris in 1897, sixty crates of specimens were delivered to Lord Rothschild at Tring: 3,075 birds, 400 birds' eggs, 150 iguanas, 65 tortoises, 40 tortoise eggs, 13 seals, 8 turtles, several hundred lizards and several hundred other zoological items (Mearns and Mearns, 1998).

A notice of the publication of Tristram's Catalogue appeared in *Ibis* (1890, 121–2) which briefly reviewed his travels and important aspects of his skin collection. The writer commented: "Not many ornithologists of the present day have enjoyed equal opportunities of studying their favourite subject in the field or have done it with such good results. Not only as an observer and collector, but also as a chronicler of his notes and discoveries, few of us, indeed, can compete with Canon Tristram."

Recognised at the present time are eight species named by him, seven species named in his honour and seventy sub-species named either by him or in his honour. Several of these are mentioned in the text or illustrated earlier but two other important species, The Duchess Lorikeet (Plate 73) and the Bare-legged Scops Owl (Plate 74), together with the sub-species Tristram's (White-collared) Kingfisher (Plate 75), were illustrated in Tristram's papers and are shown here. Tristram's collection of skins is housed in the Liverpool World Museum where they will remain actively conserved in perpetuity.

Table 11: Second Tristram Collection, Philadelphia Museum 1896–1906

| Country of Origin | No. | Collectors |
| --- | --- | --- |
| Afghanistan | 2 | H. T. Holland |
| Africa | 1 | |
| Algeria | 9 | H. B. Tristram, J. H. Gurney, H. Gillespie, N. T. Holland |
| Argentina | 5 | |
| Australia | 280 | A. S. Meek, T. C. Eyton, W. Hornby, Dunbar, Harper, J. Callinforth, Spalding, E. & C. Hose, W. H. Chamberlain, C. Holst |
| Bermuda | 3 | H. B. Tristram |
| Bolivia | 2 | |
| Borneo | 306 | J. B. Bell, A. Everett, C. Hose, T. May, Whitehead |
| Brazil | 122 | O. Salvin, A. R. Wallace, J. Collinforth, H. Whitely, C. F. Underwood |
| Burma | 129 | T. A. Hawkswell, E. W. Oates, C. T. Bingham, W. Davison, W. Harrison |
| Canada | 195 | J. K. Lord, W. Gardner, B. Field, Lt. Col. Wedderburn, L. M. Turner, H. Buckley, A. Stobart, M. Christy |
| Canary Islands | 32 | H. B. Tristram, E. G. Meade-Waldo, Koenig |

| Country of Origin | No. | Collectors |
|---|---|---|
| Caroline Islands | 2 | Finsch |
| Celebe Islands | 119 | E. & C. Hose, A. Everett |
| Chatham Island | 11 | H. O. Forbes |
| Chile | 26 | Bridges, Warwick, Tucker, Claude, E. Reed |
| China | 36 | R. Swinkoe, Severtzon, R. Lwinhoe |
| Christmas Island | 14 | A. Murray |
| Cocos Islands | 1 | W. Davison |
| Colombia | | H. Whitely, A. Arce, C. F. Underwood, O. Salvin |
| Cook Islands | 2 | |
| Costa Rica | 1296 | C. F. Underwood, H. Whitely, O. Salvin |
| Cuba | 1 | |
| Cyprus | 4 | |
| Ecuador | 8 | H. Whitely |
| Egypt | 8 | J. H. Gurney, A. L. Adams, W. C. P. Medlycott |
| England | 205 | H. B. Tristram, A. L. Chapman, J. H. Jennet, J. Callingford, J. H. Gurney, W. Procter, W. W. Saunders, R. C. Fleming, W. C, R. Burdon, J. Hunt, Lord Lilford, A. F. Sealy, T. Lourie |
| Ethiopia | 6 | Ehruberg, M. Jesse |
| Faeroe Islands | 4 | J. Chapman |
| Falkland Islands | 28 | J. Campbell, P. Thompson |
| Fiji | 33 | E. L. Layard, N. Chamberlain, D. R. Smith, Swayne, Calvert |
| France | 6 | J. H. Gurney, W. W. Saunders |
| Gabon | 4 | R. B. Walker, Hay, R. H. Beck, Hull, Harris, Hall, Bauer |
| Galapagos Islands | 10 | R. H. Beck, R. B. N. Walker, Hay, Hull, Harris, Hall, D. Bauer |
| Gambia | 6 | Dr. P. Randall |
| Germany | 7 | L. Gatke, H. Seebohm, Ussher, Swanzy |
| Ghana | 8 | Fraser |
| Greece | 2 | H. B. Tristram |
| Guatemala | 34 | O. Salvin, C. F. Umderwood, Fraser |
| Guyana | 716 | H. Whitely, C. F. Underwood |

| Country of Origin | No. | Collectors |
|---|---|---|
| Honduras | 47 | O. Salvin, G. H. W., C. F. Underwood, H. Whitely |
| Hungary | 1 | |
| India | 514 | J. G. Gillespie, J. McGregor, C. Anderson, Capt. Boys, Miss Baker, M. T. Holland, W. E. Brook, Mandelli, Miss Cockburn, T. B. Hauxwell, W. Davison, C. T. Bingham, W. W. Saunders, S. Daig, Legge, H. Whitely, Hodgson, A. Hume, J. Tweedale, J. Collingford |
| Indonesia | 4 | A. Everett, Waterstradt |
| Israel | 4 | H. B. Tristram |
| Italy | 3 | J. J. S. Whitaker, V. A. Brooke |
| Jamaica | 5 | |
| Japan | 135 | Henson, Thompson, Darcy, M. Namite, Severtzin, Pryor, Blakiston, H. Whitely, Oroston |
| Java | 12 | A. Everett, H. O. Forbush |
| Kenya | 47 | H. Whitely, N. P. F. |
| Korea | 1 | |
| Laysan | 1 | H. C. Palmer |
| Lesser Antilles | 57 | P. Gellneau, J. J. Wells |
| Lesser Sundas | 44 | A. Everett, A. R. Wallace |
| Madagascar | 773 | C. D. Cowan, J. Wills, H. W. Whitely, Gerrard |
| Madeira | 3 | Fiere, Ogilvie Grant |
| Malawi | 75 | A. Whyte, Ogilvie Grant |
| Malaysia | 93 | Singapore Museum, E. Oates, E. & C. Hose, Waterstradt, W. Chamberlain, A. Everett, C. Hose, A. C. Buller, H. E. Dresser |
| Malta | 1 | A. L. Adams |
| Marqesa Islands | 1 | |
| Mexico | 110 | A. C. Buller, H. E. Dresser, G. F. Gaumu, O. Salvin, L. Belding, C. F. Underwood |
| Moluccas | 99 | Waterstradt, A. Everett, A. R. Wallace, E. & C. Hose |
| Morocco | 3 | E. G. Heath, E. Meade Waldo |
| Namibia | 2 | C. J. Anderson |
| Nepal | 10 | Hodgson, N. W. Saunders |

| Country of Origin | No. | Collectors |
|---|---|---|
| Netherlands | 8 | J. H. Gurney |
| New Caledonia | 12 | E. L. Layard |
| New Guinea | 201 | H. B. Tristram, R. H. Richard, M. Ruggardh, A. S. Anthony, J. B. Brown, A. A. Briuju, A. R. Wallace, A. S. Meek, A. Brudy, A. Goude, A. Lanthaby, V. Rosenberg, Watersedae, Hoedt |
| New Hebrides | 1 | E. L. Layard |
| New Zealand | 151 | J. Backhouse, H. O. Forbes, H. H. Travess, H. Snow |
| Nigeria | 23 | W. Boller, A. R. Wallace, Capt. Stanley, Verreaux, J. Campbell |
| No Data | 304 | J. H. Gorney, N. Severtzon, Lord Lilford, H. Whitely, A. Everett, A. Salvin, A. Chapman, J. Callingford, J. B. Bell, Capt. Carpenter, A. B. Brooke, Miss Palmer, W. W. Saunders, E. & C. Hose, H. Cunning, Blyth |
| Norfolk Islands | 35 | P. H. Metcalf |
| Norway | 4 | H. B. Tristram, A. Pike |
| Pakestan | 10 | H. T. Holland, A. Home, W. E. Brooks, S. Ding |
| Palestine | 11 | H. B. Tristram, H. C. Fox, H. H. Jones |
| Panama | 3 | Bridges, H. Whitely |
| Peru | 97 | C. F. Underwood, H. Whitely, J. Kalinowski, E. Bartlett |
| Philippine Islands | 71 | E. & C. Hose, Waterstradt, A. Everett, H. Cummings, Cummings Collection |
| Romania | 1 | |
| Russia | 303 | Severtzon, P. A. Holst |
| Ryuku Islands | 1 | |
| St Helena | 1 | Capt. Oliver |
| Samoa Island | 3 | S. J. Whitmer, J. Brown |
| Saudi Arabia | 1 | H. B. Tristram |
| Scotland | 20 | H. B. Tristram, A. Chapman, J. Callingford, R. A. Fleming, H. M. Drummond, Small |
| Senegal | 1 | |
| Senegambia | 2 | P. Rendall |
| Seychelle Islands | 3 | H. W. Warry |

| Country of Origin | No. | Collectors |
|---|---|---|
| Singapore | 4 | W. Davidson |
| Society Islands | 3 | A. Garrett |
| Socotra | 35 | E. R. Bennett, H. O. Forbes, Forbes & Grant |
| Soloman Islands | 8 | C. M. Woodford, Rev. Welchman, A. S. Meek |
| South Africa | 234 | A. Everett, W. Bros, T. Ayres, J. H. Gurney, A. Smith, T. Atmore, C. J. Anderson, Fraser, H. Botterell, P. Rendall, G. F. S. , J. M. Benedito |
| Spain | 37 | Lord Lilford, W. W. Saunders, A. Chapman, Klase |
| Sri Lanka | 33 | W. V. Legge, T. A. Haukwell, H. Cumming, Mendelli |
| Sumatra | 17 | A. Everett, H. O. Forbes, F. Nicholson |
| Surinam | 1 | |
| Sweden | 5 | H. B. Tristram, Kjarbolling, Wheelwright, P. A. Holsp, J. Kirk, R. Swinhoe |
| Switzerland | 4 | H. B. Tristram |
| Syria | 4 | H. B. Tristram |
| Taiwan | 8 | R. Swinhoe |
| Tanzania | 1 | J. Kirk |
| Tailand | 1 | |
| Tobago | 1 | |
| Trinidad | 14 | W. L. Holland |
| Tunisia | 1 | |
| Turkey | 7 | H. B. Tristram, P. A. Holst, W. D. Cummins, K. Roper |
| Uganda | 1 | |
| USA | 71 | C. Almond, R. B. Herron, F. Stephens, E. M. Hasbronck, C. L. McKay, W. B. Holbrook, Hardy, J. G. Bell, G. M. Stoney, C. H. Townsend, Bridges, H. W. Elliot, Miss Palwer |
| Venezuela | 44 | H. Whitely |
| West Africa | 10 | Gerrard, Higgins, T. C. Eyton |
| Zaire | 1 | |
| Zambia | 19 | A. White. |
| Zimbabwe | 9 | A. White, Woodward |

### Table 12: Tristram's Birds

A. New species described by Tristram

| Species | Scientific name | Tristram's scientific name |
|---|---|---|
| **Vanuatu Scrubfowl** | *Megapodus layardi* Tristram 1879 | *Megapodus layardi* Tristram 1879 |
| **Duchess Lorikeet** | *Charmosyna margarethae* Tristram1879 | *Charmosyna margarethae* Tristram 1879 |
| **Bare-legged Scops Owl** | *Otus insularis* (Tristram 1880) | *Gymnoscops insularis* Tristram 1880 |
| **Grey-throated White-eye** | *Zosterops rendovae* Tristram 1882 | *Zosterops rendovae* Tristram 1882 |
| **Upcher's Warbler** | *Hippolais languida* | *Sylvia upcheri* Tristram 1869 |
| **Cyprus Warbler** | *Sylvia melanothorax* Tristram 1872 | *Sylvia melanothorax* Tristram 1872 |
| **Tristram's Warbler** | *Sylvia deserticola* Tristram 1859 | *Sylvia deserticola* Tristram 1872 |
| **Dead Sea Sparrow** | *Passer moabiticus* Tristram 1864 | *Passer moabiticus* Tristram 1864 |

B. New species named in honour of Tristram

| Species | Scientific name | Tristram's scientific name |
|---|---|---|
| **Tristram's Storm Petrel** | *Oceanodroma tristrami* Salvin 1896 | *Oceanodroma tristrami* Salvin 1896 |
| **Tristram's Wheatear** | *Oenanthus moesta* Lichtenstein 1823 | *Saxicola Philophthalma* Tristram 1859 |
| **Tristram's (San Cristobal) Flowerpecker** | *Dicaeum tristrami* Sharpe 1884 | *Dicaeum tristrami* Sharpe 1884 |
| **Tristram's Honeyeater** | *Myzomela tristrami* Ramsey 1881 | *Myzomela tristreami* Ramsey 1881 |

| Species | Scientific name | Tristram's scientific name |
|---|---|---|
| Tristram's Grackle | *Onychognathus tristrami* Sclater 1858 | *Onychognathus tristrami* Sclater 1858 |
| Tristram's Bunting | *Emberiza tristrami* Swinhoe 1870 | *Emberiza tristrami* Swinhoe 1870 |
| Tristram's Serin | *Serinus syriacus* Bonaparte 1850 | *Serinus syriacus* Bonaparte 1850 |

C. New sub-species described by or named in honour of Tristram

| Sub-species | Scientific name | Tristram's scientific name |
|---|---|---|
| Red-legged Partridge | *Alectoris rufa australis* (Tristram 1889) | *Caccabis rufa* var. *australis* Tristram 1889 |
| Purple Swamphen | *Porphyrio p. aneiteumensis* Tristram 1876 | *Porphyrio aneiteumenis* Tristram 1876 |
| New Zealand Snipe | *Coenocorypha aucklandica huegeli* (Tristram 1893) | *Gallinago huegeli* Tristram 1893 |
| White-throated Pigeon | *Columba vitiensis leopoldi* (Tristram 1879) | *Ianthoenos leopoldi* Tristram 1879 |
| Mackinlay's Cuckoo-Dove | *Macropygia mackinlayi arossi* Tristram 1879 | *Macropygia arossi* Tristram 1879 |
| Pinon Imperial Pigeon | *Ducula pinon salvadorii* (Tristram 1881) | *Carpophaga salvadorii* Tristram 1881 |
| Finsch's Pygmy-Rarrot | *Micropsitta finschii nanina* (Tristram 1891) | *Nasiterna nanina* Tristram 1879 |
| Nubian Nightjar | *Caprimulgus nubicus tamaricis* Tristram 1864 | *Caprimulgus tamaricis* Tristram 1864 |
| Little Kingfisher | *Alcedo pusilla richardsi* (Tristram 1882) | *Alcyone richardsi* Tristram 1882 |
| Variable (Dwarf) Kingfisher | *Alcedo lepida gentianus* (Tristram 1879) | *Ceyx gentiana* Tristram 1879 |
| Sacred Kingfisher | *Todiramphus sanctus norfolkiensis* (T. 1885) | *Halcyon norfolkiensis* Tristram 1885 |
| Tristram's (White-collared) Kingfisher | *Todiramphus chloris tristrami* (Layard 1880) | *Halcyon tristrami* Layard 1880 |

| Sub-species | Scientific name | Tristram's scientific name |
|---|---|---|
| Common Paradise Kingfisher | *Tanysiptera galatea rosselina* Tristram 1889 | *Tanysiptera rosselina* Tristram 1889 |
| White-bellied Woodpecker | *Dryocopus javensis richardsi* Tristram 1879 | *Dryocopus richardsii* Tristram 1879 |
| Short-toed Lark | *Calandrella brachydactyla hermonensis* Tristram 1864 | *Calandrella hermonensis* Tristram 1864 |
| Red-capped Lark | *Calandrella cinerea anderssoni* (Tristram 1869) | *Megalophorus anderssoni* Tristram 1869 |
| (Common) Crested Lark | *Galerida cristata macrorhyncha* Tristram 1859 | *Galerida macrorhynca* Tristram 1859 |
| (Common) Crested Lark | *Galerida cristata arenicola* Tristram 1859 | *Galerida arenicola* Tristram 1859 |
| Melanesian Cuckooshrike | *Coracina caledonica welchmani* (Tristram 1891) | *Graucalus (Artimedes) welchmani* Tristram 1891 |
| Melanesian Cuckooshrike | *Coracina caledonica lifuensis* (Tristram 1879) | *Graucalus lifuensis* Tristram 1879 |
| Yellow-eyed Cuckooshrike | *Coracina lineata nigrifrons* (Tristram 1892) | *Graucalus nigrifrons* Tristram 1892 |
| Slender-billed Greybird | *Coracina tenuirostris salomonis* (Tristram 1879) | *Edoliisoma salamonis* Tristram 1879 |
| Black-browed Triller | *Lalage leucopyga affinis* (Tristram 1879) | *Symmorphus (Lalage) affinis* Tristram 1879 |
| White-throated Dipper | Cinclus cinclus minor Tristram 1870 | *Cinclus minor* Tristram 1870 |
| Island Thrush | *Turdus poliocephalus samoensis* Tristram 1879 | *Turdus samoensis* Tristram 1879 |
| Mourning Wheatear | *Oenanthe lugens halophila* (Tristram 1859) | *Saxicola halophila* Tristram 1859 |
| Desert Wheatear | *Oenanthe deserti homochroa* (Tristram 1859) | *Saxicola homochroa* Tristram 1859 |
| Capped Wheatear | *Oenanthe pileata livingstoni* (Tristram 1867) | *Campicola livingstoni* Tristram 1867 |
| White-headed Blackchat | *Mymecocichla arnoti arnoti* (Tristram 1869) | *Saxicola arnoti* Tristram 1869 |
| Black-chested Prinia | *Prinia flavicans ortleppi* (Tristram 1869) | *Drymoeca ortleppi* Tristram 1869 |

| Sub-species | Scientific name | Tristram's scientific name |
|---|---|---|
| Cetti's Warbler | *Cettia cettia orientalis* Tristram 1867 | *Cettia (Potamodus) orientalis* Tristram 1867 |
| Polynesian Reed Warbler | *Acrocephalus aequinoctialis pistor* Tristram 1885 | *Acrocephalus pistor* Tristram 1885 |
| Long-billed Reed Warbler | *Acrocephalus caffer mendanae* Tristram 1883 | *Acrocephalus mendanae* Tristram 1883 |
| Solomon (Islands) Pied Monarch | *Monarcha barbata vidua* (Tristram 1879) | *Piezorhynchus vidua* Tristram 1873 |
| Solomons Flycatcher | *Monarcha barbata squamulatus* (T. 1882) | *Piezorhynchus squamulatus* Tristram 1882 |
| San Cristobal Myiagra | *Myiagra ferrocyanea cervinicauda* Tristram 1879 | *Myiagra cervinicauda* Tristram |
| Rufous Fantail | *Rhipidura rufifrons russata* Tristram 1873 | *Rhipidura russata* Tristram 1873 |
| (Common) Golden Whistler | *Pachycephala pectoralis christophori* Tristram 1879 | *Pachycephala christophori* Tristram 1879 |
| Black Sunbird | *Necterinia serisea christianae* (Tristram 1889) | *Cynnyris christianae* Tristram 1889 |
| European Chaffinch | *Fringilla coelebs palmae* Tristram 1859 | *Fringilla palmae* Tristram 1859 |

— 16 —

# Ootheca Tristramiana: Collector to Conservationist

OSPREY. (*Pandion haliaëtus.*)

Tristram and Alfred Newton were close friends and there can be little doubt that together they discussed the planning of many, if not all, of the expeditions that the former undertook. Another friend of both was John Wolley who made probably the most famous egg collection of all time (Newton, 1864). Tristram would have learned a great deal about oology from both of these friends. As Newton published papers he sent copies of them to Tristram who eventually had them bound up as a single volume, which many years ago was acquired by the present writer. It carries the bookplates of both Tristram and W. H. Mullens, the well-known ornithological bibliographer, and one of the papers it contains is entitled 'Suggestions for forming a Collection of Birds' Eggs' (Newton, 1860), an activity now frowned upon and illegal. The concluding observations of this paper, reprinted with additions from a circular prepared by Newton for the Smithsonian Institution of Washington, read as follows: "The best allies of the collector are the residents of the country, whether aboriginal or settlers, and with them he (the collector) should always endeavour to cultivate a close intimacy, which may be assisted by the offer of small rewards for the discovery of nests and eggs."

Tristram was acting in this way long before the publication of the pamphlet in either the Smithsonian Institution or in Great Britain and had done so successfully since his time in Bermuda and later in Algeria and Palestine. It is quite probable that Newton got this idea from Tristram. By the time the latter sold his collection in his later years, he had amassed some 24,000 eggs of 2,120 species. In addition to these numbers he had sold and exchanged many others, for example his "triplicate eggs" from Algeria. These brought £250 at Stevens' Sale Room in Covent Garden in 1858, the year after he had returned from the Great Sahara. Single eggs of Golden Eagle sold for £1 12s. 0d. (£1.60p) and a particularly well-marked single egg for £2 15s. 0d. (£2.75p), whilst Kentish Plovers' eggs were sold at between 8s. and 11s. (40p and 55p) and Little Egrets' between 14s. and 18s. (70p and 90p) each. Some of the prices for which single eggs were sold far exceeded those of the Wirral, Cheshire dealer C. H. Gowland in 1947, when it was still legal to sell them. Working to an eight-hour day, 24,000 eggs would take almost a year to prepare for the cabinet. Even if Tristram collected only half of these it would have occupied a great deal of his time as he had only occasional help from an assistant.

Special mention is given in the Collection Leaflet, which Tristram prepared for Thomas Parkin (Parkin. ca. 1900a), to three items in his possession of which he was particularly fond. Firstly, his Great Auk egg (Plate 76)—which was also a great rarity—as by the time that he acquired it the species was extinct. He paid the taxidermist J. Williams of Oxford Street, London, £30 for it in 1853, according to the Collectors' leaflet, but £35 pounds according to Errol Fuller in his superb account of the Great Auk (Fuller, 1999). This particular egg had a somewhat chequered history and it is thought to have been collected in Iceland, probably on Eldey in 1834 (Newton, A., MS., Cambridge University Library). It was probably amongst one of the last batches of eggs of the species taken to Copenhagen. When Tristram sold his egg collection to Philip Crowley for £1,000, the Great Auk egg was included in the purchase and it eventually was taken into the British Museum collection, now at Tring, in Hertfordshire, when they accepted those eggs which were of use to them from this larger collection.

Tristram's egg is not a particularly handsome one, being of a common variety with a chalky white background, with small spots and occasional scrolling, the markings being evenly spaced across the surface. Whilst most eggs of the Great Auk are of a similar shape to those of the Common Guillemot, this egg tends in shape towards that of the Razorbill.

Secondly, Tristram drew attention to a single egg of the Great Bustard which he possessed and which is thought to be the last laid in England by the native stock. The egg was taken by Mr J. D. Salmon: "In the spring of 1832 three females resorted to Great Massingham Heath, in Norfolk, for incubation. Their eggs consisted of two pairs and a single one. These were taken away, under the impression that as there was no male bird, they were good for nothing" [but] "the male is said to live apart after the female is impregnated". (Salmon, J. D., 1833) This single egg was the one which entered Tristram's collection and was purchased by him for the sum of 27s. (£1.35p) at the sale of the collection formed by Mr T. C. Heysham.

Whilst in the possession of Mr H. D. Salmon the egg was borrowed by William C. Hewitson for illustration in the latter's *Coloured Illustrations of the Eggs of British Birds* (3rd edn, Plate LXXIII). In the text accompanying the plate of the egg, Hewitson (1856) writes: "The increasing taste for ornithology, and the mania to possess a British-killed specimen of any bird, has made it hopeless that any such bird, especially of large size, should long escape with life having unluckily landed on our shores."

Thirdly, Tristram draws attention to a single egg of the Greenshank in his collection, "the first authentic egg of the species known to Ornithologists", taken by William MacGillivray on the island of Harris: "The Greenshank [Plate 77], whose nest had never been found before in Britain, we detected breeding in various parts of the country, generally in some swampy marsh, or by the margin of some of its numerous lakes." (MacGillivray, 1852)

It might be thought that Tristram, as an ornithologist, would have gained more satisfaction from finding for himself the nests of species previously unknown. This he did during his visit to Norway and Finnmark, when in 1852 he discovered for the first time the nests of the Great Snipe (Plate 78), Bar-tailed Godwit (Plate 79) and Green Sandpiper. Whilst the eggs of the Great Snipe were not exactly unknown at the time of Tristram's visit to Norway, none of the eggs thought to be of this species was properly authenticated. In the first two editions of Hewitson (1831, Plate CLIII; 1846, Plate LXXXVI), an illustration of an egg is included that is said to be that of a Great Snipe, sent to Hewitson by a Mr Hoy. It is almost certainly of that species. However, in the third edition of Hewitson (1856) this plate is omitted and two eggs collected by Tristram in 1852 are substituted. Tristram is quoted: "I found the bird breeding in great numbers in marshy swamps near Bodo, in Nordland, in the early part of the summer of 1852; I shot several birds from the nest." In the following August, Tristram shot several more in the same area, a marsh near Bosoe—"we obtained, I think, seven braces in half an hour" (Tristram, 1853c).

Surprisingly, John Wolley (Plate 80), a bird-nester and oologist par excellence, appears not to have found the nest of the Great Snipe in the long periods he spent in Scandinavia and all the eggs of that species in his collection (Newton, 1864–1907), twenty-two in all, were purchased or collected by other workers. Number 4257 in his collection was obtained from Tristram in 1854: "Mr Tristram gave me this egg at Castle Eden, 25 August 1854 and wrote upon it in my presence." The two eggs illustrated in the third edition of Hewitson (1856) were selected as typical from a series of fourteen in Tristram's collection. All the nests that Tristram found near Bodo contained clutches of four eggs. The single egg given to Wolley was from a clutch of four in Tristram's cabinet, so diminishing that clutch.

"Eggs said to be of the Bar-tailed Godwit were brought from the north by the Rev. H. B. Tristram and also by Mr Wolley with evidence sufficient to justify me figuring them. The Bar-tailed Godwit was seen in close vicinity to the eggs, the Black-tailed Godwit was never seen at all." (Hewitson, 1856) Tristram himself comments: "I found the bird in the breeding season in Finnmark, and shot several specimens in breeding plumage, where I found no trace whatever of the black tail; I got the nest and shot the bird, a female, the same morning close to the spot but I did not flush her off the nest." Tristram compared the eggs he had taken with those that Wolley had collected and neither could distinguish between them. Tristram observed "that they were at once to be recognised among twenty eggs of the common godwit". Hewitson records that he was unable to make the distinction but Tristram clearly had a sharper eye. On average the egg of the Bar-tailed Godwit is smaller, lighter in weight and more plover-like in shape, whereas the eggs of the Black-tailed Godwit are more snipe-like in shape. It is fair to say that Tristram could legitimately claim to have collected the first authentic eggs of the Bar-tailed Godwit to be brought back to the British Isles.

Lastly, Tristram is rightly credited with finding the first properly authenticated eggs of the Green Sandpiper and these are illustrated by Hewitson (1856). In Tristram's egg catalogue they are listed under "578 Totanus ochropus. Taken by myself in Nordland, Norway, Offoden Fjord in July 1852. All three eggs are . . . from different nests." One nest had three eggs and the other two each had a clutch of four. Tristram found the best areas for both Green Sandpiper and Bar-tailed Godwit were over into Sweden, in the provinces of Tornea and Lulea. In addition to the egg of the Great Snipe which Tristram gave to Wolley from his Norwegian trip he gave him a single egg of the Green Sandpiper (Newton, 1864). "Mr Tristram told me that this was one of those taken by him in Norway in 1852 of which some were sold at Stevens' saleroom, 9th May 1854." It is probable that either Wolley or Tristram was mistaken in this since it was blown with two holes and was probably obtained from a dealer. Six Green

Sandpiper eggs were sold in the Stevens' sale. Wolley's comments about the sale are unclear as he suggests that Tristram both bought and sold Green Sandpipers' eggs on this occasion.

Many of the eggs in Tristram's collection were singles and in many cases it seems that he took only one egg from a clutch, in contrast with most of the later collectors who preferred to exhibit full clutches in their cabinets. Many of the earlier collectors such as Tristram recorded details of their collections in notebooks whereas later collectors tended to use data cards kept with the particular egg(s), though some used both methods and others merely wrote on the eggs. These different methods often provided difficulties, even up to the present day, when catalogues become separated from collections with the result that the data may be lost for all time, or in cases of litigation that the data cards are being deliberately withheld. The separation of the Catalogue and the eggs unfortunately happened in the case of Tristram's collection after it was sold to Mr Philip Crowley in 1887, when the Catalogue formed part of the sale. On its acquisition, Crowley began a new 'Catalogue of Birds' Eggs' in seven quarto volumes bound in half morocco. Volume 1 was labelled 'British' and the other six 'General'. The content of the Tristram 'Egg Registers' (Plate 81) was incorporated into the New Catalogue which was in Crowley's hand. In 1896 Crowley began another New Catalogue which was unfinished in four quarto volumes at the time of his death in 1897, the handwriting in this being in the hand of Crowley's secretary. Philip Crowley's collection at this time was the largest in private hands, he having incorporated that of his brother Alfred and the whole of Tristram's collection; it contained the eggs of over four thousand species.

By his will, Crowley gave the Trustees of the British Museum the power to take from the cabinets "all such specimens as would be useful for making their series more complete." However, they were not offered his Catalogues or Tristram's Listings at this stage but Tristram had written data on many eggs which were subsequently transcribed to data labels in the museum. For many years Tristram's Listings were separate from that part of his collection housed in the British Museum and it is possible that had the museum been offered the Listings at the time of its acquisition they may have taken many more specimens.

The eggs not taken from the collection by the museum authorities went for auction at Stevens' Auction Room in 1902, requiring a three day sale on 17 April, 15 May and 6 June. Many were purchased by Thomas Parkin, of Hastings, who had been elected to the British Ornithologists' Union in 1880, and was known to Tristram. Parkin was the compiler of the leaflets on *Well-known Collectors and their Collections* from which that of Philip Crowley (Parkin, ca. 1900b) provides much of the information given here (In Tristram's Egg Registers, British Museum, Tring).

After Crowley's death in 1901 and the acquisition of part of Tristram's egg collection by Parkin, the new owner exhibited a series of the eggs in the Hastings Museum. In a printed leaflet, a copy of which is present in the Tristram Egg Registers in the Natural History Museum at Tring, the history is outlined of the egg of a Sooty Tern (quite a rarity at this time) which was exhibited. The egg, No. 2448, was collected by John James Audubon on the Tortuga Keys, USA, in May 1832. Audubon gave this specimen to Prof. William MacGillivray of Aberdeen University. After the latter's death Tristram acquired his collection, not from the widow as stated by Parkin, but from some other member of the family, as MacGillivray's wife pre-deceased him. The egg of the Sooty Tern was purchased by Rowland Ward at Stevens' Sale Room in 1902 and purchased from him by Parkin in 1904. Parkin commented in the leaflet describing the exhibition, that Sooty Terns lay only one egg because "the power of producing them is almost exhausted" because of continuous robbery of the eggs. However, a single egg is the normal clutch size, though occasionally clutches of up to three eggs do occur.

The Egg Catalogues and Registers associated with the Philip Crowley collection passed into the hands of his nephew, Mr Reginald Crowley, of Croydon, who was himself a member of the BOU. He loaned them to Thomas Parkin who had purchased many of the eggs. Somehow or other, probably from Reginald Crowley, Parkin acquired the 'Tristram Registers' and in his time loaned them to the British Museum. This is evidenced by a letter, dated 1 July 1912, which like other material is now bound into those

Registers, now in Tring. This letter from Thomas Wells thanks Parkin for lending the volumes to Mr Ogilvie Grant and is a cover note returning them. The Tristram Registers remained with Thomas Parkin until he made a gift of them to the museum on 13 April 1927 when they were eventually re-united with that part of Tristram's collection retained by the museum. Before this catalogue was initiated by Tristram he used a copy of one of the rare printed catalogues designed by the Rev. S. C. Malan which is still in existence (Williams, K. F., 2010). Here Tristram's Great Auk egg is recorded "330a. Bought from Williams, Oxford St., for £30, 1853".

The collection of eggs formed by Tristram was exceptional by any standards. Much of what he collected was new to science, so much so that it is not possible to discuss it here; it would be a book in itself. A few examples have been selected for special mention, particularly Tristram's favourites from his collection and a few highlights of his discoveries. From over 24,000 specimens it is difficult to make such a choice. Tristram's was the age of collecting, describing and classifying, all fundamental to the development of biological sciences. Nowadays such collecting is frowned upon and it is only in exceptional circumstances, and under licence, that eggs or birds are collected, and this, usually, only for scientific purposes. We have learned a great deal from the first real ornithologists, of which Tristram was one but early in the twentieth century and before the death of Tristram an influential section of the British Ornithologists' Union questioned the continuing validity of oology—egg collecting. The argument was put that little if anything was to be gained by taking birds' eggs that could not be gained by observation. This section of the BOU became openly antagonistic towards oology and as a reaction to this an 'Oological Section' of the Union was formed in 1911. The oologists exhibited eggs at meetings of the British Ornithologists' Club, which was formed in 1892 "facilitating the social intercourse of the BOU". Members of the Club had to be members of the Union. Later, in 1922, the antagonism to egg collecting was apparent in the BOC and this led to the formation of the British Oological Association, as the BOC ceased to publish proceedings of the Oological Section. The British Oological Association was later to become the Jourdain Society, in memory of F. C. R. Jourdain, one of the original editors of the *Handbook of British Birds* (Witherby, H. F., Jourdain, F. C. R., Ticehurst, N. F. and Tucker, B. W., 1938), but the Society has been recently dissolved and the Jourdain Society National Collection of Birds' Eggs moved to the Natural History Museum at Tring, in Hertfordshire. The naming of the now defunct Society after him was done in recognition of his contribution both in terms of his writing, e.g., most of the breeding biology in the 1938 Handbook and his contribution to what was considered to be the science of oology. At the outbreak of the Second World War a friend asked Jourdain's daughter what her father might do. "Bury his Great Auk egg" was her reply (Spurling, 1995).

Much of the evolving thought leading to the sidelining of the oologists and the formation of the Jourdain Society was prompted by the increasing influence of the Royal Society for the Protection of Birds. Eventually, through the Protection of Birds Acts, 1954–67 their influence led to the death knell of private, legal egg collecting. This did continue illegally, but now it is generally dying out and egg collections are almost exclusively limited to museums.

Tristram was throughout his life at the forefront of ornithological change and development, from the formation of the British Ornithologists' Union, through the publication of its journal, the *Ibis*, the organisation of the sort of expeditions which had not previously been known, to the beginnings of the conservation movement, of which he was one of the first conservationists. To a large extent many people began to object to the wholesale slaughter of birds during the breeding season (which was legal at the time) by so-called 'sportsmen'. Historically, the cry was taken up by Professor Alfred Newton (who else?) at a meeting of the Section of Zoology and Botany at the British Association Meeting held in November 1868, which was widely commented on by journals and the press. In 1869 the Seabirds Protection Bill was passed in Parliament and the BA appointed a committee for the purpose of investigating the desirability of establishing a 'close time' for the preservation of indigenous animals.

This committee was initially presided over by Tristram, and its most prominent members were Newton and another oologist, J. E. Harting. They were frequently called upon to give advice to the Committees of the House of Commons and the first 'success' was the 1872 Bill for the Protection of Wildfowl. Newton regarded this as an unmitigated disaster as it was an ineffective compromise. The development of the law on Bird Protection, the initiation of the conservation movement and the part played by these three leading oologists will be dealt with in the next chapter.

Tristram was influential at all levels from national government to local society and he can truly be considered as a founder member of the conservation movement. But like Newton he never dropped his interest in birds' eggs. It is indeed ironical that the protection movement was initially affected so much by three such oologists as Newton, Tristram and Harting. But it was, and no one can deny their genuine interest in ornithology, their love of birds and their great wish that all birds be protected. Were it not for these men and their great influence at all levels, it may well have been many years before public opinion had had a similar effect on legislation. Tristram was a collector of birds and eggs, which was what ornithologists did then, but he was one of the very first people to propagate the drive for the protection of birds and the conservation of their habitats.

Plate 76: Tristram's egg of the Great Auk *Pinguinus impennis*. Reproduced with the permission of the British Museum (Nat. Hist.).

Plate 77: Eggs of the Greenshank *Tringa nebularia*. The eggs of this species were first taken by William MacGillivray on the island of Harris, Outer Hebrides, and were at one time in Tristram's collection but are now in the Natural History Museum, Tring. The clutch illustrated is from the collection of the Liverpool World Museum.

Plate 78: Eggs of the Great Snipe *Gallinago media*. The eggs of this species were first discovered by Tristram on his expedition to northern Scandinavia in 1852. The clutch illustrated is from the collection of the Liverpool World Museum.

Plate 79: Eggs of the Bar-tailed Godwit *Limosa lapponica*. The eggs of this species were first discovered by Tristram on his expedition to northern Scandinavia in 1852. The clutch illustrated is from the collection of the Liverpool World Museum.

Plate 80: In Memory of John Wolley, this plaque was placed in Southwell Cathedral, Nottinghamshire.

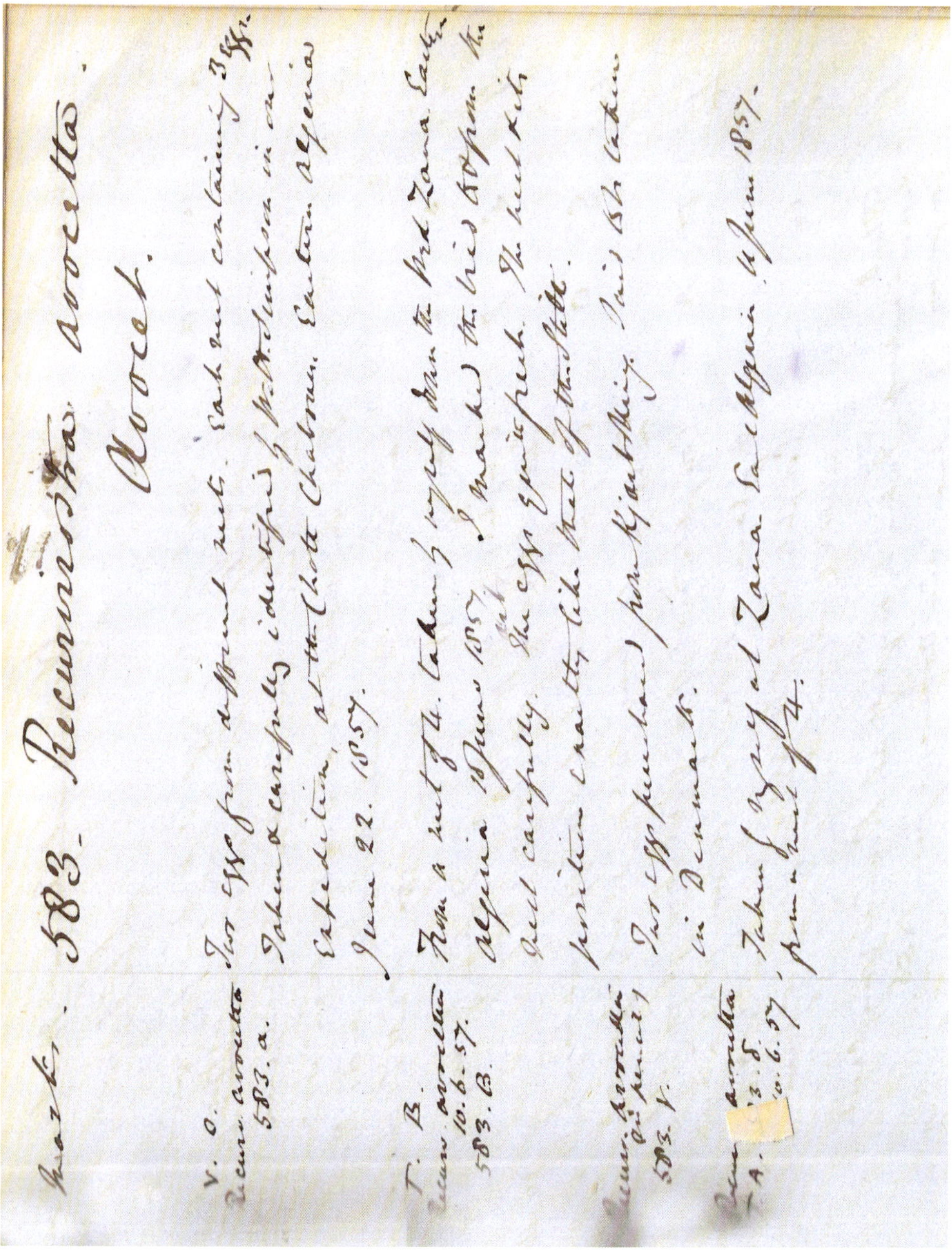

## 583. Recurvirostra Avocetta. Avocet.

**V. O.**
*Recurv. Avocetta*
583. α.

Ten eggs from different nests. Each nest containing 3 eggs. Taken & carefully identified by Mr Thompson in an expedition at the (Salt) Chotts, Eastern Algeria. June 22. 1857.

**T. B.**
*Recurv. avocetta*
10. 6. 57.
583. β.

From a nest of 4. taken by myself near the Petit Zana Lake, Algeria 10 June 1857. I marked the bird both in the eat. She sat equivalently. The nest was well carefully. position exactly like that of the shelldr.

*Recurv. Avocetta*
(0. 6. 1857)
583. γ.

Ten eggs purchased from Kjaerbölling in Denmark.

*Recurv. avocetta*
A. 3. δ.
10. 6. 57.

Taken by myself at Zana Lake Algeria 14 June 1857. from a nest of 4.

Plate 81: A page from Tristram's Egg Catalogue.

Plate 82: Nesting Seabirds on the Farne Islands. Shooting such birds in large numbers resulted in the first legislation for their protection, the Seabirds Preservation Act 1869.

Plate 83: Kittiwake *Rissa tridactyla*. The slaughter of large numbers of this species during the breeding season, for "sport" and for their feathers, resulted in a public outcry.

# REPORT

FROM THE

## SELECT COMMITTEE

ON

# WILD BIRDS PROTECTION;

TOGETHER WITH THE

PROCEEDINGS OF THE COMMITTEE,

MINUTES OF EVIDENCE,

AND APPENDIX.

*Ordered, by* The House of Commons, *to be Printed,*
23 *July* 1873.

Plate 84: Title page from the report of the Select Committee.

# MINUTES OF EVIDENCE.

*Thursday, 12th June 1873.*

MEMBERS PRESENT:

Mr. Dillwyn.
Mr. Hambro.
Mr. Heron.
Mr. Auberon Herbert.
Mr. Andrew Johnston.

Mr. Jones.
Colonel Parker.
Mr. Rowland Winn.
Sir David Wedderburn.

THE RIGHT HONOURABLE E. W. M. AUBERON HERBERT, IN THE CHAIR.

The Reverend HENRY BAKER TRISTRAM, F.R.S, called in; and Examined.

*Chairman.*

1. You were Chairman of the Committee of the British Association which recommended the preparation of the Bill which was introduced by Mr. Andrew Johnston last year, were you not?—I was.

2. The Bill was for the protection of certain edible birds, was it not?—It was for the protection of all web-footed and wading birds, including plover, snipe, woodcock, sand-pipers, gotwit, and all the goose and duck tribe, including also the rails, most of which are good for human food.

3. Are you yourself prepared to see the protection which you wish to give to those birds extended to other classes of birds which would have been unprotected under that Bill?—Perfectly so; but I should like to explain that I think they stand on a different footing; my impression, as a practical naturalist, having devoted 25 years study to the subject, is that all marsh birds, and water fowl, are rapidly diminishing in England; I say nothing about the west of Scotland, which I do not know well, but in England, and in the east and lowlands of Scotland they are rapidly diminishing, not only from the drainage, for that has not so much affected them, except as circumscribing their breeding localities, and making them more exposed to interruptions, but because, owing to the price which those birds will fetch when there is no other game in the market, it is well worth the while of the poacher to kill them on the nest for the supply of Leadenhall. In consequence of those birds being præternaturally tame at that time of the year, they are becoming rapidly extinct. With regard to those that are already exterminated, or that are in course of being exterminated, I think it is most necessary that whatever protection is afforded to them should be

*Chairman—continued.*

hedged about by sufficient penalties against the man who makes his money by killing them, and that the sale should be prohibited under any circumstances. I have gone into the question at Leadenhall; I find that the exemption on birds that profess to come from abroad is futile; that home birds, and birds that I know have been killed in England, are sent up to Leadenhall Market as Dutch birds. So long as they are allowed to sell birds of those species during close time they will never be protected; those are the large birds which are articles of human food. Then again with regard to small birds, I do not think that they are decreasing in the same proportion, on account of the decrease of their natural enemies, and owing to the increase of population and of game preserving; my own recollection of the wildest country districts of Northumberland, Durham, Roxburghshire, and Yorkshire, is that the small birds, since I was a boy and took birds' nests, have very much increased, and that most of our small insectivorous birds have increased. I believe that, with the exception of a single species, here and there, there are, generally speaking, as many small birds in the country as would naturally find food; any decline with regard to small birds has not been on account of their being shot. I do not think that they are shot during the breeding season; it is not worth while; they can take care of themselves. Their real danger is from bird-catchers. Although I would not prohibit the bird-catcher's calling at proper seasons, it is most unfair that he should be allowed to catch them during the breeding season. I think that a law which prohibited bird-trapping, say between the 1st of April and the 1st of August, on the highway, or within 50 yards

Rev. *H. Tristram* F.R.S.

12 Jun 1872.

0.100.  A  of

Plate 86: "Climmers" collecting eggs on the cliffs at Bempton. Image by permission of Jim Whitaker.

— 1 7 —

# Bird Protection and the Grassroots of Conservation

STORK. (*Ciconia alba.*)

The Great Auk was gone—dead as the Dodo—and even the ever-optimistic Newton had to admit it. The fact that it had been alive during the lifetimes of both Newton and Tristram had, to some extent, turned their thoughts from the creation of new species to how the loss of more of the old ones might be prevented. Thus it was that at the 1868 meeting of the British Association held in Norwich, Newton delivered a paper on 'The Zoological Aspects of the Game Laws' (Newton, 1869). Newton's biographer claimed that "he clearly showed that the wholesale slaughter of birds during the breeding season would shortly result in their extinction unless laws were passed to give them protection." (Wollaston, 1921) Many, then and now, would question this statement, as Newton certainly would, but Newton was aware that the general public had an increasing desire to protect our wild animals. At the following meeting of the BA, in Exeter, Newton's paper was followed up by Tristram, who delivered a paper entitled 'On the effect of Legislation on the Extinction of Animals'. Again Newton and Tristram were thinking along the same lines and working in concert.

Ancient laws had protected Herons for the entertainment of falconers, and Partridges and Pheasants for the 'sportsmen' who shot them. However, there was no protection for the Robin or Blackbird in the garden or the Kittiwake on her nest on the cliff face (Plate 82). Particularly during the breeding season birds are more vulnerable than at other times of the year and shooting the Kittiwake could well cause her young to die. In his talk Newton particularly called for the protection of the birds of prey and seabirds during the breeding season (Plate 83) and drew attention to the seasonal slaughter of seabirds particularly on the Isle of Wight and Flamborough Head. Surprisingly there was widespread publicity of the contents of Newton's paper and public condemnation for those people living in Bridlington, the town nearest to the cliffs of Flamborough Head.

As a result, the Rev. Fredrick Barnes-Lawrence, the vicar of the Priory Church at Bridlington, called a meeting of local naturalists and clergymen to consider the matter. The group established the Association for the Protection of Seabirds which was supported by the local Member of Parliament, Christopher Sykes; the Archbishop of York, William Thompson; and the author of the then popular bird book *A History of British Birds* (1851-7) Francis Orpen Morris.

Local naturalists circulated pamphlets and argued against the shooting on moral grounds. However, the local boatmen increased their income significantly by taking shooting parties to the cliffs at Bempton, and other people came long distances to watch, so that the local economy received a significant boost. Kittiwake feathers were collected for the millinery trade and moral indignation was not sufficient to stop the killing of literally thousands of birds. Newton, in his talk to the BA, quoted one person who claimed to have "in one year killed with his own gun four thousand (Kittiwakes) ... and I was told of another of these sea-fowl shooters who had an order from a London House for ten thousand." He further argued for legislation to bring in a 'close time' to extend over the breeding season (Newton, 1868). There was close co-operation between Newton and the Association for the Protection of Seabirds and this extended to the co-operation with the committee set up in August 1868, at the Norwich meeting of the BA, to look into the practicability of establishing a 'close time' for the protection of indigenous animals. Initially Newton was not the chairman of this group but provided evidence to its first meeting chaired by Tristram who had also taken an interest in the matter, particularly since it applied initially to the North-East.

The first meeting of the 'Close Time' Committee took place in the Zoological Society rooms on 13 January 1869. The Committee consisted of Tristram (Chairman), Frank Buckland, N. B. Tegetmeier and H. E. Dresser. At the first meeting J. E. Harting was added to the membership and Alfred Newton attended. The Report for 1869 (BA, 1870–80), apart from being a record of the meeting, contained information on the progress and eventual passing of the Seabirds Preservation Act 1869.

Mr C. Sykes, M.P., progressed the Bill through the Commons and His Grace the Duke of Northumberland through the Lords "where it met with a most favourable reception". Probably due to the involvement of Tristram and Newton with the Yorkshire Association for the Protection of Seabirds, the information which Mr Sykes put to the Commons was of a very positive nature, avoiding the moral issues and considering only the practical reasons for protecting seabirds. The main reasons given were that seabirds were of use in destroying grubs and worms, in acting as scavengers in the harbours, in warning vessels off the rocks by their cries during fogs, and in hovering over and pointing out to the fishermen the locality of the shoals of fish.

Initially the Bill had been drafted for a close season from 1 March to 1 August but during its passage through the House this was changed from 1 April to 1 August. On the argument that it was traditional for those living close to the sea to gather eggs and young birds as food, egg protection and the protection of young birds before they could fly were also dropped from the Bill. It was also considered that it was expedient to exempt the island of St Kilda, the inhabitants of that island being so dependent on seabirds for their subsistence. So it was that the Bill passed through both Houses of Parliament surprisingly quickly and even Newton, who was always critical of anything done by administrators, particularly if they knew nothing of the subject concerned, was reasonably pleased.

There is interesting wording in the final draft of the Bill. Presumably to ensure that the likely contraveners of the new law knew which birds were protected the local names of each species were included in the listing. Thus Bonxie (Great Skua), Coulterneb (Puffin), Loon (any Diver), Marrot (Razorbill or Puffin), Murre (Guillemot or Puffin), Scout (Razorbill, Guillemot or Puffin), Seamew (Common Gull), Sea-parrot (Puffin), Sea-swallow (any Tern or Storm Petrel), Solan Goose (Gannet), Tarroch (Kittiwake—immature or any Tern or Guillemot), Tystie (Black Guillemot) and Willoch (Common Guillemot or Puffin) appeared together with the more usual common names of each species, which are here bracketed after each local name. Thus, the Puffin occurs six times in the list but was not any better protected for it.

All this was reported at the first meeting of the British Association 'Close Time' Committee and the rest of the meeting was devoted to consideration of the benefits that birds provided. Newton's evidence was most convincing. His knowledge of the continental literature in addition to that of the Americans and, of course our own, seemed to result in the Committee finding evidence for benefits exceeding depredations in all cases. The Committee considered that there were two problem areas, one being agriculture where birds were thought to take too many seeds and the other the 'game preserver' who consider all raptors as vermin. In this report the

Committee made the statement that "the best method of judging of the good or harm done by birds is that of studying the nature of their food" and they were anxious to provide objective evidence for their claims. Parliament had clearly been impressed by the data supposedly demonstrating that more ships had been wrecked on Flamborough Head during the period when shooting was allowed compared with a similar period when it was outlawed, and also the claim that seabird calls helped to direct ships away from the nesting cliffs. In both cases a simple statistical test would probably have demonstrated that the data did not support the claim.

Two years earlier, at Dundee, Tristram had suggested that the Grouse epidemic was the result of the indiscriminate slaughter of predatory animals. He claimed that birds of prey are "the sanitary police of nature", and that if they had existed in their original strength they would have stamped out the Grouse disease, in as much as hawks (and Falcons) in preference make sickly animals their quarry (Tristram, 1868b). In the unlikely situation of this being demonstrable the possibility of Parliament taking any action in favour of the predators, or any action of any sort based on evidence provided as it turned out, was something the Committee was to find in the coming years.

The first Report of the 'Close Time' Committee to the BA ended with three main comments. Firstly, "Much information as to the nature of the food of birds is, however, yet needed in order to judge correctly of the amount of good or harm they do"; secondly, "We would suggest that the best method of affording the necessary protection of birds would be to prohibit the carrying of a gun, during the breeding season, as is done in several parts of the continent" and, thirdly, "Much information is, however, yet needed as to the practical working of the 'close time' system in those countries in which it has been in force."

On the whole, at the close of 1869, the 'Close Time' Committee members were pleased with the progress that they had made and the two papers by Newton and Tristram had set the Protection ball rolling. The first Bill had passed through Parliament but little did they know that they would meet for another eight years before more progress was made.

At the second meeting of the Committee, towards the end of 1870, the membership had changed somewhat. No comment on this is made in the text of the Report which lists Alfred Newton, H. B. Tristram, J. E. Harting, H. Barnes and H. E. Dresser as the members present. Tristram remained Chairman (BA Report, 1873).

Little progress had been made with the establishment of a 'close time' because of Parliamentary difficulties but they were able to report that where the Seabirds Act was enforced very beneficial results followed. In the hope that similar results might be obtained with the extension of protection to other groups of birds, they suggested that "wildfowl" might be considered for protection, particularly since there was a considerable reduction in the number of plovers and sandpipers and very great numbers of ducks (geese, too, are probably considered under this heading) were also shot in spring.

The 1871 Report by the Committee, with the same membership as in 1870, was even shorter, and stated that an Association had been established whose members were pledged not to destroy Woodcock during the breeding season. The Committee proposed that in order to prevent a further decline in numbers of plovers, by shooting and trapping during the breeding season, a move might be made for a Wildfowl Bill.

The Report of 1872, again by the Committee constituted as in the previous year, confirmed that as a result of modifications in Parliament a Bill for a 'close time' for all birds was approved by the Commons, after much alteration of the original draft and an initial rejection by the House of Lords. Item 7 of the BA Report reads: "Your Committee cannot but look with mixed favour on this measure. It appears to them to attempt to do too much and not to provide the effectual means of doing it." Further it "is mischievous in its effect since it diverts public attention from those species ... which, through neglect, indifference, caution, cupidity or prejudice are suffering a persecution."

As a result of the widening of the Bill the possibility of a child being prosecuted for throwing a stone at a Robin, to quote a simple example, resulted in much lower penalties being included. Following on from this there was little or no reduction in the number of wilfdfowl in the spring markets; the penalties had no effect. The Parliamentary Select Committee, numbering twenty-three members, contained no one who could be counted an ornithologist and no ornithologist was consulted. The Bill generally regarded as useless by the Committee was passed in an almost deserted House by thirty votes to fifteen. In a letter to his brother Edward, Alfred Newton wrote: "Mr Herbert, on the 21st of June last, laid a cuckoo's egg in the carefully built nest of the BA Committee, and the product is a useless monster—the wonder alike of the learned and the layman, and an awful warning as an example of an amateur legislation." To Tristram he wrote: "All this we owe to the fools of enthusiasts. The Wildfowl Bill, followed next year by one for the regulation of bird catchers would have done far more good." (Letter in Wollaston, 1921 dated 29.7.1872) More formally, in the Report for 1873 to the BA, the 'Close Time' Committee stated initially that the Wild Bird Protection Act "gave almost universal discontent". Three members (not named) of the 'Close Time' Committee gave evidence—which was apparently ignored—to the Select Committee and as a result (Item 5), "Your Committee wish emphatically to condemn these recommendations as a whole, and all but one of them separately." The three-page Report merely lists the Committee's objections to the 1872 Wild Birds Protection Act.

Completely separate from the activities of the BA Committee, a Report (Plate 84) was published on 23 July 1873 from 'The Select Committee of Wild Birds Protection together with the Proceedings of the Committee, Minutes of Evidence and Appendix'. In fact there were fifteen Appendices and a thirty-nine page Index, including an Analysis of the Index, the whole running to 255 folio pages (Select Committee, 1873). This included details of eight meetings between 20 May and 23 July, the last being on the date of the publication of the Report. This Report was probably generated by complaints from many sources concerning the content of the 1872 Wild Birds Protection Act. As a result the Select Committee on Wild Birds' Protection decided to interview a cross-section of the public and thirty-nine men were called as witnesses.

Nowhere in the Report is it made clear how the witnesses were selected but the list included a cross-section of those in the community with an interest in birds and it was composed of five clergymen, five gardeners, five farmers, an architect, a bookseller, a banker, a tradesman, a picture dealer, a journeyman tailor, a militiaman, an army officer, the Secretary of the RSPCA, two bird-catchers, the Assistant Keeper of Zoology at the British Museum, several Justices of the Peace, several who had written books on birds, members of Naturalists' Societies, Alfred Newton—Professor of Zoology and Comparative Anatomy in the University of Cambridge, and Canon Henry Baker Tristram—Canon of the Cathedral of Durham. Surprisingly there were no significant landowners apart from Lord Lilford. Both the last mentioned and the bird-catchers were supportive of full protection during the breeding season, as were the rest of the group, though several suggested that a list of exceptions to the full protection be made.

The first witness to be called was the Rev. Henry Baker Tristram, F.R.S., who it was stated was the Chairman of the Committee of the British Association which recommended the 1872 Bill put forward to the House of Commons by Mr Andrew Johnson. The minutes of the evidence ran to 157 pages, 12 of them being contributed by Tristram, (Plate 85) significantly more than any other witness.

Much of the evidence given by witnesses was anecdotal and some of the more serious aspects not properly understood by the examining committee. What really ought to have been obvious to anyone was that fines for breaking the law concerning shooting ought to be of sufficient magnitude to discourage it. This was clearly not appreciated by the examining committee. Both Tristram and Newton separately stressed this part of their evidence, the former on 12 June, the latter on 19 June. The point was also mentioned by several other witnesses and Newton stressed it in a separate paper, six folio pages in length, submitted to the Select Committee (Appendix 10 of the Report of the Select Committee).

Much was made of those birds regarded as beneficial to humans, those which were edible and those threatened with extinction. Clearly, many of the witnesses were not satisfied with what had been discussed during their questioning as is shown by the fifteen appendices, all printed in full in the Report, which were submitted later.

At the meeting on 23 July (Report of the Parliamentary Select Committee on Wildbird Protection, 1873) six RESOLUTIONS were decided upon:

1. That the protection of certain wild birds named in the Schedule of the Wild Birds Protection Act of 1872 be continued.
2. That all other wild birds be protected from 15th March to 1st August, provided that the owners or occupiers of lands, and persons deputed by them, have permission to destroy such birds on lands owned or occupied by them.
3. That one of Her Majesty's Principal Secretaries of State should be empowered to except, in any particular district, any bird from the protection afforded, either by the Act of 1872 or by the proposed Act, if he think necessary to do so.
4. That for the sake of giving better protection to the swimmers and waders, no dead bird, if such bird is mentioned in the Sea Fowl Preservation Act, or the Wild Bird Protection Act of 1872, be allowed, from 15th March to 1st August, to be bought or sold, or exposed for sale, whether taken in this country or said to be imported from any other country.
5. That no bird-catcher at any time of the year be allowed to set his nets or snares of any description within 50 yards of the high road, or of any park or garden, unless by permission of the owner or occupier of such park or garden.
6. That any violation of this proposed Act, or of the Wild Bird Protection Act of 1872, be punished by the payment of costs alone for the first offence, and the payment of costs, and a fine not exceeding 5s. for every offence after the first.

Resolution 5 effectively rendered the first four useless as the fine was no deterrent compared with the profit made in breaking the law. The inconsistencies of the Select Committee were not limited to its proposals; for the three days' expenses allowed to witnesses Tristram received £7 7s. 0d. travelling from Hartlepool whereas Mr Scot-Skirving, travelling from Chester received £9 12s. 0d.

The Report of the 'Close Time' Committee for 1874 stated that in the House of Lords Earl De la Warr introduced a Bill entitled 'An Act for the more effectual Protection of Wild Birds in the Breeding Season', adding that "The Bill is not based on any of the recommendations of the Select Committee of the House of Commons (1873) and still less on any suggestions which have proceeded from your Committee." The Bill was withdrawn. The criticism of this Bill was mainly associated with "rendering penal the taking of certain bird's eggs, and the firm belief of the 'Close Time' Committee that the only effectual protection which was practicable was the protection of the adults during the breeding season."

The 1875 Report of the 'Close Time' Committee was very short; nothing had been achieved concerning the introduction of a Bill into Parliament but Mr Henry Chaplin, M.P. for Mid-Lincolnshire "holds out the hope that he will at an early period of the next session bring forward such a measure." This he did, and the Bill passed through the Commons, the only modification being the shortening of the close season. In the Upper House Lord Henniker took charge of the Bill which received Royal Assent on 24 July 1876. The 1876 Report called attention to the very satisfactory working of the Seabirds Preservation Act which was a result of higher penalties for its contravention. This Report also contained a summary of the previous Reports of the 'Close Time' Committee "in the hope that Parliament might consider the implementation of measures previously ignored."

The 1877 Report was short and commented only on the shortcomings of the 1876 Bill and the fact that several counties had made application to the Home Office for alteration in the 'close time' to bring it more into line with those originally recommended by the 'Close Time' Committee.

The Reports for 1878 and 1879 were noted only in the British Association Section for Zoology and Botany. It appears from the Reports that the 'Close Time' Committee met over a period of nine years and the results were negligible. This was through no fault of their own, but largely through the ignorance of the Parliamentary Select Committee and their failure to take note of the measures recommended to them. Parliament seemed to respond more to public opinion than it did to its Select Committee and there was some justification in this. The climate of opinion was in favour of more protection for wildlife in general and for measures to be taken against egg collecting.

Both Tristram and Newton would have been disappointed to have achieved so little with so much effort involved over an eleven-year period.

The 'Close Time' Committee was stood down. Throughout the life of the Committee Tristram and Newton had opposed any legislation against egg collecting, firstly, on the grounds that much of this was carried out by schoolboys who it would be unreasonable to prosecute and, secondly, because they both considered that it would have little effect. Both wanted to concentrate on protecting the adult birds during the breeding season and Newton particularly was anxious to set up areas which were totally protected against shooting and egg collecting during the breeding season.

In 1880 the Select Committee was left with the summary of eleven years of advice from the 'Close Time' Committee, and the Wild Birds Protection Act 1880 was introduced. This included a close time for all wild birds between 1 March and 1 August, with a list of exceptions. The Act was better framed but its scheduled birds were not properly protected and within a few years many clauses were amended. The Act was effectively unworkable. Within the Act local councils could apply for an order to protect species locally and in several counties this worked well. Despite difficulties in bringing prosecutions, interpreting the law and the continued depredations of collectors, the 1880 Act moved things forward and even Newton admitted that there was progress. In 1893 Sir Herbert Maxwell introduced a Bill to Parliament which aimed to allow County Councils to prohibit the taking of birds' eggs. This passed through the Commons but fell foul of the Lords and was dropped. The identification of birds' eggs during prosecutions made it difficult to apply laws and only blanket protection would have avoided these problems.

The biggest step forward in the field of bird protection in the British Isles came in 1889 with the formation of the Society for the Protection of Birds and the first subscription of one guinea was subscribed by Professor Alfred Newton. Even though the 1894 Act was drafted by the Society, Newton regarded it as imperfect as those of 1880 and 1896. The whole situation was summed up in the *Historical Atlas*: "These Acts provided for the protection of known species of birds during the breeding season, different for each county, but it was often said that it was 'nobody's business' to put the Acts into motion against the offenders." (Holloway, 1996)

Throughout the development of protection methods for wild birds in Britain, both Tristram and Newton put in a lot of time and effort to promote such measures, though they opposed specific action against the taking of eggs. Both preferred protection for eggs to be through the establishment of areas where collecting was forbidden. Both thought it reasonable to allow schoolboys to take eggs, for eggs to be taken for food in some circumstances and the 'scientific' collection of birds' eggs to be legal. They opposed large-scale dealing in eggs and the wholesale collection for profit of any form (Plate 86). Together they began the measures for protection of birds but they died, Tristram in 1906 and Newton in 1907, almost fifty years before a satisfactory Act was implemented in 1954. Tristram and Newton usually acted together in anything of a scientific nature and, in fact, in anything to do with birds. Towards the end of their lives they worked together on aspects of bird protection and the beginnings of the conservation movement. Together they were the first to take action on behalf of wild birds and should be recognised for it.

In addition to being chairman of the BA 'Close Time' Committee for eleven years, Tristram was a member of the Association for the Protection of Seabirds, based in Bridlington, between 1868 and 1874; of the Protection of Wild Birds' Eggs Committee of the BA between 1891 and 1894; of the Yorkshire Naturalists' Union Wild Birds' Eggs Committee between 1894 and 1897, and Vice-President of the Society for the Protection of Birds between 1892 and 1904, and then of the Royal Society for the Protection of Birds as it became in 1904 until his death in 1906. Newton declined such a position. Had the legislation of the 1954 Protection of Birds Act been implemented during the period when Tristram and Newton were still alive it is possible that both would have been amongst the first to be prosecuted, despite all their good work for the protection of birds and the conservation movement. Could they really have given up egg collecting? It would certainly have been another quandary for Tristram.

On 17 October 1896 the University of St Andrews awarded Tristram an Honorary Doctorate "for his life-long services to ornithological science and literature". In 1875 the Dean and Chapter of the Cathedral had relinquished its responsibility of the river banks in Durham to Tristram and these duties he took very seriously. Here was his own small nature reserve and it is appropriate that his last publication (Tristram, 1906) was a letter to the Durham County Advertiser calling for greater protection of Tawny Owls along the banks of the river Wear within the city.

— 18 —

# Last Letters

TAWNY OWL. (*Syrnium aluco.*)

For over fifty years Tristram and Newton had frequently corresponded with each other despite the fact that they met often at scientific meetings and visited each other in Durham and Cambridge. Many of these letters survive in private collections and particularly in the Cambridge University Library (9839/1T/201–259). Newton annotated and kept most of those which he received but Tristram was not so systematic, though he often tipped in to books in his library correspondence which he received from their authors. The sale Catalogue of his books refers to several of these but the location of most of them is now unknown. Wollaston (1921) refers to many of them but does not reveal their location at the time of his writing, though it is possible to find some of them in the Cambridge University Library. It is mainly from these letters that it is possible to trace aspects of Tristram's life for he kept no detailed record except when he was on his various expeditions and here the records are in his papers and books. His correspondence is therefore of some importance historically.

Tristram died on 8 March 1906, aged 83. The final letters between these close friends are both interesting and touching. From his sick bed Tristram wrote to Newton (Wollaston, 1921):

> The College, Durham.
> February 25, 1906.
>
> My dear Newton,
>
> It is utterly impossible to get out of your debt epistolary, as I have found ever since that act of friendship many years ago, when you took off your name from the Royal Society in order to secure my election. When one looks back through the long vista of years there is nothing I have found to equal it for self-sacrifice and generosity. But, that apart, there is only one sense of generous fraternity. I am glad to deliver my soul.
>
> In my present state of health the political outlook hardly interests me, for I am very ill. The doctor stays in the house with me generally all night, as I suffer from breathlessness, but I have had the comfort of having my family, eight children and my sons-in-law gathered round me yesterday.
>
> Before my attack became worse, I was able to enjoy two papers in the Ibis on Ross's Rosy Gull and the Scotch Antarctic, as well as to glance at my old friend Mr Whittaker's Tunisia, which resuscitated my interest in bygone years. I dare say this may be my last letter.
>
> Your sincere old friend,
> H. B. Tristram

Newton received this letter the following day and replied immediately (Wollaston, 1921):

> Magdalene College, Cambridge.
> February 26, 1906.
>
> My dear Tristram,
>
> If the letter I have just received is really to be the last I am to have from you, as therein foreshadowed, there could not be one more gratifying to my feelings, and I am at a loss for terms in which to answer it. I can never forget the steady, friendly, I may say brotherly support I have invariably received from you, and if it were my good fortune to have done you a good turn in the matter of the Royal Society, a circumstance that had wholly passed from my mind, it was but a slight return for the aid you rendered in starting the BOU and the Ibis, and again at the critical moment when our first editor threw up the job, and (with one or two more) would not have been sorry if it had come to an end. It was your Palestine papers that to a very great extent caused the success of the second series of the Ibis; not that I would overlook the value of the help I had from Blyth, Swinhoe and others. Their articles were of great scientific interest, but they failed in the qualities for reading, while your articles possessed both merits.
>
> There are few things I look back on with greater satisfaction than my six years editorship of the Ibis, but that was entirely due to my contributors, among whom you were chief, and one on whom I could always rely.
>
> I can't trust myself to write more, and indeed you might easily be tired with more. It must indeed have been a comfort to you to have all your children round you on Saturday, and so I say farewell, and God be with you till you are with Him.
>
> Yours as always,
> Alfred Newton.

In his last letter Tristram refers to reading on his death bed J. I. S. Whitaker's book *The Birds of Tunisia* (1905). This contains excellent plates by Gronvald (reproduced here as Plates 31 to 34) of Tristram's north African Larks and Chats, and other birds which he saw and collected on his expeditions—his reference here to "my interests in bygone years".

It was on the third anniversary of his wife's death that Tristram died and he was buried *super montibus sanctis* ("upon the holy hills")—part of the motto of the University of Durham—beside the great Cathedral facing Palace Green. Sacred Ibis was no more and the Great Gun of Durham would not be heard again. Just over a year later, Newton died on 7 June 1907. One of the great ornithological partnerships was no more. I would like to think that they had both found the Great Auk but perhaps that is too much to ask.

— 1 9 —

# Tristram's Library

Tristram's library was extensive, much of it having been inherited from his father (Tristram, L. H-H., 1898). His daughter also commented that Tristram himself was a great bibliophile and his writings confirm that he had access to an extensive literature. There can be little doubt that his friendship with Newton afforded access to the Cambridge University libraries and Newton's own and that Newton's access and knowledge of the foreign literature was a great help to him. However, more concrete evidence of his own collection occurs in the form of a sale 'Catalogue of Books, a Portion of the Library of the late Rev. Henry Baker Tristram, LL.D. offered for sale by Sotheby, Wilkinson and Hodge, on 12 June 1908.' (Plate 87)

As the cover of the Catalogue indicates the sale was of only a small part of Tristram's library. Of books relating to birds there were some 205 titles in 154 lots. Some lots contained several titles but there were enough to give a general impression of the library as a whole. Of the important periodicals which contained ornithological material there were full runs of:

- *Ibis* (50 volumes)
- *Proceedings of the Zoological Society* (90 volumes), plus the five volumes of its predecessor the *Zoological Journal*
- *Transactions of the Zoological Society* (18 volumes)

There were shorter runs of other journals:

- *Asiatic Journal*, 1816–19; vols I–VIII
- *Bulletin de la Société Zoologique de France*, 1876–85, vols I–X
- *Degland et Gerbe Ornithologie Européenne*, 1867, 2 vols in 4
- *Magazine of Natural History*, 1829–36
- *Natural History Review*, 1861–5
- *Natural Science*, 1892–96, vols 1–9
- *New Zealand Institute, Transactions and Proceedings*, 1869–1903, vols I–XXXV
- *Revue Africaine*, 1856–68, vols I–XII
- *Revue Zoologique*, 1845–8, 4 volumes

General Reference books included works on Africa, Asia, Australia (signed and presentation copies from John Gould), North America and India, together with a small selection of County Avifaunas. Several volumes concerned ornithological exploration, for example Gray's and Sharpe's *The Zoology of the Voyage of Erebus and Terror* and 'museum' collections. There was a full run of the twenty-seven volumes of the *Catalogue of Birds in the British Museum*, a work which Newton frequently criticised but is considered by some to be a "fine bird book" (see below), and the *Hand-list of Genera and Species of Birds in the British Museum*. Both these in my own library proved invaluable in translating Tristram's scientific names into those used at the present time. There were several works on the oology of different countries and what are now termed "fine bird books" by antiquarian book dealers and collectors. This term refers mainly to fine illustrations, usually hand-coloured, but many of the works are extremely valuable from both the historic and scientific points of view. The sale Catalogue included:

- Albin, E., 1738–40. *Natural History of Birds.*
- Dresser, H. E., 1884–6. *Monograph of the Meropidae, or Family of Bee-eaters.*
- Gould, J., 1832. *Monograph of the Ramphastidae or Family of Toucans.*
- Gray, G. R., 1849. *The Genera of Birds.*
- Jardine, W., and Selby, P. J., 1825–43. *Illustrations of Ornithology.* AL from Selby inserted.
- Mivert, S., 1896. *Monograph of the Lories or Brush-tongued Parrots.*
- Rothschild, W., 1893–1900. *The Avifauna of Laysan and the neighbouring Islands.*
- Sclater, P. L., 1882. *Monograph of the Jacamars and Puff-Birds.*
- Sharpe, R. B. and Wyatt, C. W., 1885–94. *Monograph of the Hirundinidae or Family of Swallows.* AL from author inserted.
- Sharpe, R. B., 1868–71. *Monograph of the Alcedinidae, or Family of Kingfishers.* ALS from Sharpe inserted.
- Shelley, G. E., 1876–80. *Monograph of the Nectarinidae or Family of Sunbirds.*
- Shelley, G. E., 1896–1905. *The Birds of Africa, comprising all the species which occur in the Ethiopean Region.* 4 vols. In 5.

Most of Tristram's books on British ornithology had obviously gone elsewhere though the family do not appear to have kept anything of an ornithological nature and all that remain in their possession are copies of the books written by Tristram. Many other books from Tristram's library found their way into the collections of later ornithologists and I have copies which were later in the libraries of W. H. Mullens and Abel Chapman. The latter is a presentation copy of Seebohm's *Geographical Distribution of the Charadriidae* which Tristram received from Seebohm in 1887, and includes a letter to Tristram from a former Palestinian dragoman David Jamal. Surprisingly it carries a date of 1899 under Chapman's signature. This perhaps suggests that Tristram was disposing of some of his books before his death. What is also surprising is that this presentation copy to Tristram is the five guineas plate edition as Seebohm's usual gift of this work was the £2 12s. 6d. edition which lacked the plates. Seebohm must have held Tristram in very high regard.

Another point which might give an indication that Tristram was preparing to sell at least some of his books comes from an American paper published in 1919 by W. Foxon on the ornithology of Wilson and Bonaparte, where it is reported that a copy of the *Illustrations of Game Birds* was listed in "R. H. Porter's Catalogue of the late Rev. H. B. Tristram's Library. 1906, £3-3s, but I do not know where the copy now rests." The existence of this Catalogue suggests that there may have been a sale previous to that of Sotheby's 1908 but no trace of Porter's Catalogue has been found.

Over 4 per cent of the books in the Sotheby's sale were presentation copies and many others were signed and contained autographed letters. It is always a sad day when a library such as that of Tristram is disposed of after the death of the owner but this is almost always inevitable when there is no one in the family of similar interests to whom it might be bequeathed. Newton left his library and papers to the Zoology Department in Cambridge University where most of his books remain; manuscripts have been transferred to the University Library.

The 146 lots which could be said to contain any ornithological interest realised a total of approximately £530 in the sale on 12 June 1908. Of this total the twelve titles listed above as "fine birds books" brought in £70 1s. 0d.

Now, in a similar sale, the same twelve titles would realise between £50,000 and £60,000! However, the true value of this library was not monetary but its academic utility to its owner in the production of his long series of papers. Both Lord Rothschild and Lord Derby possessed more extensive and more expensive libraries, but Tristram's was a working library which he had built up carefully himself and was amongst the best such specialist libraries in the country at the time. For anything that he required that was missing from it he could always turn to Newton.

# CATALOGUE OF BOOKS.

### A Portion of the Library of the late
### REV. HENRY BAKER TRISTRAM, LL.D.
*Canon of Durham (deceased)*

(SOLD BY ORDER OF THE EXECUTORS).

### OCTAVO ET INFRA.

#### LOT 1.

JARDINE (Sir W.) Naturalist's Library, 40 vol. *portrait and coloured plates, half morocco, uncut, t. e. g.*    Edinb. 1833

#### 2.
Lesson (R. P.) Histoire Naturelle des Oiseaux de Paradis et des Épimaques, 40 *coloured plates, half morocco, t. e. g.*    Paris, s. d.

3   Finsch (O.) Die Papageien, 2 vol. *coloured plates, etc. half morocco t. e. g.*    Leiden, 1867-68

4   Heuglin (T. von) Ornithologie Nordost-Afrika's, de nilquellen und Küsten-gebiete, des Rothen Meeres und des Nordlichen Somal-Landes, mit Appendix, 4 vol. *fine coloured plates, half calf*    roy. 8vo.   Cassel, 1869-74

5   Hume (A.) Nests and Eggs of Indian Birds, Rough Draft (662 pp.) *half calf, presentation copy from the author to Canon Tristram*    roy. 8vo.   1873

6   BRITISH MUSEUM. CATALOGUE OF THE BIRDS IN THE BRITISH MUSEUM, by R. B. Sharpe, H. Seebohm, Hans Gadow, P. L. Sclater, etc. 27 vol. complete, 387 *fine coloured plates, original cloth,* SCARCE
   *printed by Order of the Trustees,* 1874-98

7   Jerdon (T. C.) The Birds of India, 3 vol. *autograph letter of author to Rev. H. B. Tristram enclosed*    Calcutta, 1862-64

8   Giebel (C. G.) Thesaurus Ornithologiae, 3 vol. *half morocco*    roy. 8vo.   Leipzig, 1872-77

9   Finsch (O.) und G. Hartlaub. Beitrag zur Fauna Centralpolynesiens, Ornithologie der Viti-Samoa und Tonga-Inseln, *coloured plates, half morocco, t. e. g.*    roy. 8vo.   Halle, 1867

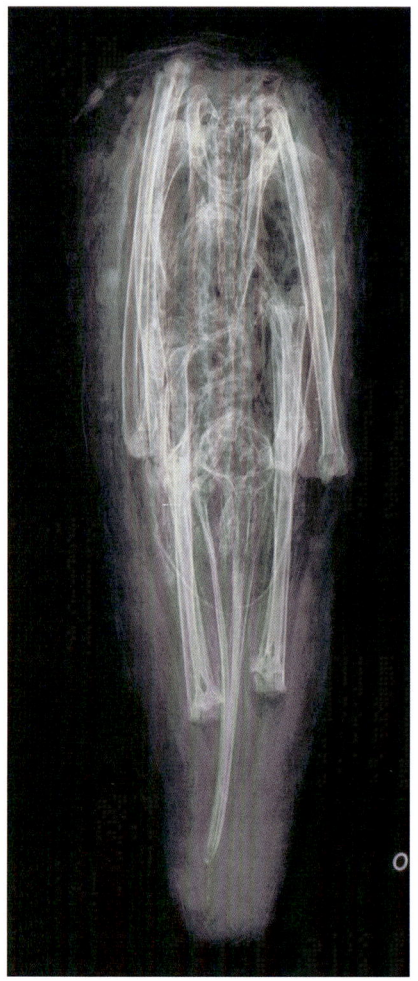
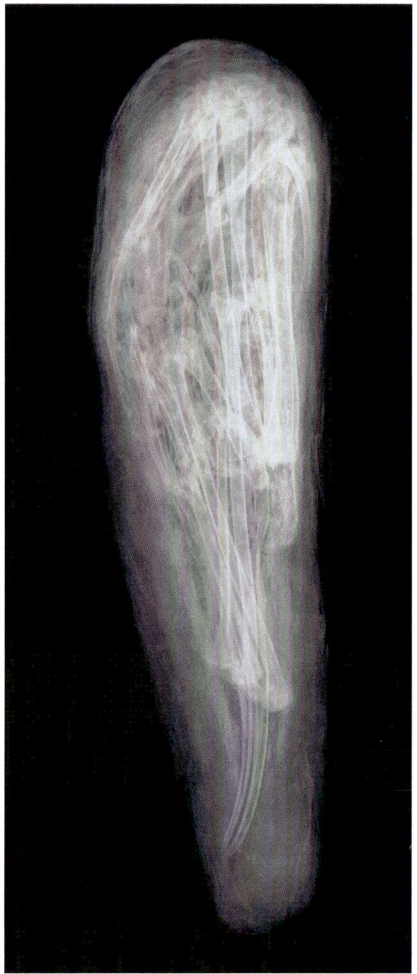

**Plate 88: Mummified Sacred Ibises. X-ray images and the mummy case.**

# References

Audubon, J. J. L., 1831–9. *Ornithological Biography*. (Adam & Charles Black).

Audubon, J. J. L., 1840–4. *The Birds of America*. (J. J. Audubon & J. B. Chevalier).

Baker, R. A., 1996. ' "The Great Gun of Durham"—Canon Henry Baker Tristram, F.R.S. (1822–1906)' in *Archives of Natural History* 23 (3):327–41.

Baker, R. A., 1999. 'Going, going, gone—the "Durham" Great Auk' in *Archives of Natural History* 26 (1):113–19.

Barrett-Hamilton, G. E. H., 1897. Letter to *Ibis*, Seventh Series 3:484–6.

Benson, A. C., 1911. 'The Leaves of the Tree—Professor Newton' in *The Cornhill Magazine*, March.

Bewick, T., 1847. *A History of British Birds*, 8th edn. (Longman & Co.).

Bodenheimer, F. S., 1956a. 'H. B. Tristram's Collections in Natural History, especially of Palestine' in *Annals of Science* 12, 4:278–87.

Bodenheimer, F. S., 1956b. 'Canon H. B. Tristram 1826–1906. A Preliminary Bibliography, Obituaries and other Biographical notes' in *Proceedings of the University of Durham Philosophical Society* XIII, Series A (Science). No. 3:12–22.

Bodenheimer, F. S., 1957. 'Canon Henry Baker Tristram of Durham (1822–1906)' in *Durham University Journal* XLIX (New Series XVIII):95–7.

Bolam, G., 1912. *The Birds of Northumberland and the Eastern Borders*. (Henry Hunter Blair).

Brazil, M. A., 1991. *The Birds of Japan*. (Smithsonian Institute Press).

Brinkley, Edward, 2012. 'Bermuda Petrel (*Pterodroma cahow*)'. Neotropical Birds On Line (T. S. Schulenberg, Editor). Ithica. Cornell Laboratory of Ornithology.

British Association for the Advancement of Science. 1870–80. *Reports (Annual) of the 'Close Time' Committee of the British Association*.

Bryson, W., 2003. *A Short History of Nearly Everything*. (Black Swan).

Chilton, Glen, 2014. *The Curse of the Labrador Duck.* (Simon and Schuster).
Darwin, C., 1859. *On the Origin of Species by Means of Natural Selection.* (John Murray).
Darwin, C., 1871. *The Descent of Man.* (John Murray).
Darwin, C., and Wallace, A. R., 1858. 'On the Tendancy of Species to form Varieties, and on the perpetuation of Varieties and Species by Natural Means of Selection' in *J.Proc.Linn.Soc.Zool.* 3:45–62.
Del Hoyo, J., Elliot, A. and Sargatal, J., (eds), 1992–2013. *Handbook of the Birds of the World.* (Lynx Editions).
Dobson, A. and Madeiros, J., 2006. 'Bermuda', in Sanders S. M., ed., *Important Bird Areas in United Kingdom Overseas Territories*, 29–36 (Royal Society for the Protection of Birds).
Dresser, H. E., 1871–90. *A History of the Birds of Europe.* (Published by the author).
Evans, A. H., 1900. *Birds.* The Cambridge Natural History, vol. IX. (MacMillan & Co.).
Fisher, J., 1953. 'The Collared Turtle Dove in Europe' in *British Birds* XLVI:151–81.
Fisher, J., 1954. *Birds as Animals. A History of Birds.* (Hutchinson).
Fisher, J. and Petersen, R.T., 1964. *The World of Birds.* (Macdonald).
Fleming, Eleanor M., n.d. 'Recollections'. (Unpublished MS., Durham University Library).
Forbes, H. O. and Robinson, H. C., 1898. *Catalogue of the Parrots (Psittaci) in the Derby Museum.* (Bulletin of the Liverpool Museums, 1:5–22).
Foreshaw, J. M. and Cooper, W. T., 1973. *Parrots of the World.* (Landsdowne).
Fuller, E., 1995. 'Great Auk good buy' in *British Birds*, 88:610.
Fuller, E., 1999. *The Great Auk.* (Published by the author).
Gaskell, J., 2000. *Who Killed the Great Auk?* (O.U.P.).
Gladstone, H. S., 1927. 'Letter "Labrador Ducks"' in *Ibis*, Series 12, 3:352–3.
Goode, G. B., 1909. 'Dr P. L. Sclater: Biographical notices of the original members of the British Ornithologists' Union' in *Ibis*, Series 9, 2:129–37.
Goodman, S. M. and Meininger, P. L., 1989. *The Birds of Egypt.* (O.U.P.).
Gould, J., 1837. 'Description of new species of Finches collected by Darwin in the Galapagos' in *Proc.Zool.Soc.Lond.* 5:4–7.

Green, V. H. H., 1996. *A New History of Christianity*. (Continuum International).

Greenway, J. C., 1958. *Extinct and Vanishing Birds of the World*. (American Committee for International Wildlife Protection).

Grieve, S., 1885. *The Great Auk, or Garefowl* (Alca impennis, *Linn*.) (Thomas Jack).

Gunther, A., 1908. 'Obituary of Canon H. B. Tristram' in *Proc.Roy.Soc.B.*, 80:xlii–xliv.

Hancock, J., 1874. *A Catalogue of the Birds of Northumberland and Durham*. (Williams and Norgate).

Harting, J. E., 1891. *Bibliotheca Accipitraria. A Catalogue of Books ancient and modern relating to Falconry*. (Quaritch).

Hewitson, W. C., 1831. *Coloured Illustrations of the Eggs of British Birds*. 1st edn., Plate LXXXVI. (Currie n.d., Bowman).

Hewitson, W. C., 1856. *Coloured Illustrations of the Eggs of British Birds*. 3rd edn., Plate LXXIII. (Van Voorst).

Hewitson, W. C., 1859. 'Recent Discoveries in European Oology' in *Ibis*, Series 1, 1:76–80.

Higgins, P. J., 1999. *Handbook of Australian, New Zealand and Antarctic Birds*. (O.U.P.).

Holling, M. et al., 2012. 'Rare Breeding Birds in the United Kingdom' in *British Birds* 105:394.

Holloway, S., 1996. *The Historical Atlas of Breeding Birds in Britain and Ireland*. (T. & A. D. Poyser).

Houlihan, P. F., 1986. *The Birds of Ancient Egypt*. (American University Cairo Press).

Irestedt, M., Fuchs, J., Jonsson, K. A., Ohlson, J. I., Pasquet, E. and Ericson, P. G. P., 2008. 'The Systematic Affinity of the enigmatic *Lamprolea victoriae* (Aves: Passeriformes)—an example of Avian Dispersal between New Guinea and Fiji over Miocene intermittent land bridges?' in *Molecular Phylogenetics and Evolution*, 48:1218–22.

Jardine, W., 1848–53. *Contributions to Ornithology for 1848–53*. 5 vols. (Edinburgh).

Jones, J. M., 1859. *The Naturalist in Bermuda*. (Reeves & Turner).

Jourdain, Silvester, 1610. *A Discovery of the Bermudas.* (University of Virginia Press).

Latham, J., 1821–8. *A General History of Birds.* 2:171. (Published by the author).

Lerner, H., Meyer, M., James, H., Hofreiter, M. and Fleischer, R., 2011. 'Multilocus Resolution of Phylogeny and Timescale in the Extant Adaptive Radiation of Hawaiian Honeycreepers' in *Current Biology* 21:1–7.

Linnaeus, C., 1758. *Systema Naturae. Regnem Animale*, 10th edn. (Stockholm).

MacGillivray, W., 1852. *A History of British Birds.* (Scott, Webster & Geary).

Mayr, E., 1951. 'Speciation in Birds' in *Proc.X Int.Orn.Congr.Uppsala*, 1950:91–131.

Mayr, E., 1966. *Animal Species and Evolution.* (Cambridge Belknap Press).

Mearns, B. and Mearns, R., 1988. *Biographies for Birdwatchers. The lives of those commemorated in Western Palaearctic bird names.* (Academic Press).

Mearns, B. and Mearns, R., 1998. *The Bird Collectors.* (Academic Press).

Meyer, H. L., 1835–42. *Illustrations of British Birds.* (Longman and Co.).

Moreau, R. E., 1959. 'The Centenarian "Ibis"' in *Ibis* 101:19–38.

Morris, F. O., 1851–7. *A History of British Birds.* (Groombridge & Sons).

Mountfort, G., 1959. 'One Hundred Years of the British Ornithologists' Union' in *Ibis* 101:8–18.

Newton, A., 1860. *Suggestions for Forming Collections of Birds' Eggs.* (Smithsonian Misc.Coll.).

Newton, A., 1861. 'Abstract of Mr. J. Wolley's researches in Iceland respecting the Gare-fowl or Great Auk' in *Ibis*, Series 1, 3:374–99.

Newton, A., 1864–1907. *Ootheca Wolleyana.* (R. H. Porter).

Newton, A., 1864. 'On the irruption of Pallas's Sand-Grouse (*Syrrhaptes paradoxus*) in 1863' in *Ibis*, Series 1, 6:185–222.

Newton, A., 1865. 'The Gare-fowl and its Historians' in *Natural History Review*, 467–88.

Newton, A., 1866. 'On some little known works which mention *A. impennis*' in *Ibis*, Series 2, 2:223–4.

Newton, A., 1869. 'The Zoological Aspects of the Game Laws. Norwich, 1868' in *British Association Report*, 108–9. Murray.

Newton, A., 1870. 'On Existing Remains of the Garefowl' in *Ibis*, Series 2, 6:256–61.

Newton, A., 1879. 'Gare-fowl' in *Encyclopaedia Brittanica*, 9th edn, 10:78–80. (A. & C. Black).

Newton, A., 1885. 'Mr. Grieve on the Gare-fowl' in *Nature* (October 8), 545–6.

Newton, A., 1888. 'The Early Days of Darwinism' in *Macmillan's Magazine*, No. 340.

Newton, A., 1896. *A Dictionary of Birds*. (A. & C. Black).

Newton, A., 1898. 'On the Orcadian home of the Gare-fowl' in *Ibis*, Series 7, 4:587–5.

Newton, I., 1985. 'Irruption', in Campbell, B. and Lack, E. A., eds, *Dictionary of Birds*. (Buteo Books).

Noble, G., 1931. *The Birds of Jesmond Dene*. (Eyre and Spottiswood).

Parkin, T., n.d. [ca. 1900a]. *Well-known Collectors and their Collections. Canon Henry Baker Tristram, D.D., F.R.S.* (Privately published Leaflet. See Tristram Egg Catalogue, Tring).

Parkin, T., n.d. [ca. 1900b]. *Well-known Collectors and their Collections: Philip Crowley.* (Privately published Leaflet).

Payne-Gallwey, G., 1886. *The Book of Duck Decoys*. (Van Voorst).

Phillips, J. C., 1922. *A Natural History of the Ducks*. (Houghton Mifflin Co.).

Purchas, Samuel, 1625. *Purchas his Pilgrimes*. (Henfie Featherstone).

Reid, S. G., 1883. 'The Birds of the Bermudas' in *Royal Gazette*. (Hamilton).

Rowley, G. D., 1875–7. *Ornithological Miscellany* 2:205–27. (R. H. Porter).

Ruse, M., 2001. *Can a Darwinian be a Christian? The Relationship between Science and Religion*. (C.U.P.).

Salmon, J. D., 1833. 'Of the Great Bustard (Otis tarda)' in *Mag.Nat.Hist.* 7:150.

Salvadori, T., 1880–82. *Ornithologi della Papuesa e delle Molocche*. (Torino).

Salvadori, T., 1891. 'Catalogue of the Psittaci, or Parrots, in the Collection of the British Museum' in *Catalogue of the Birds of the British Museum* XX:9–10. (British Museum).

Salvin, O., 1959. 'Five months Birds'-nesting in the Eastern Atlas' in *Ibis*, Series 1, 1:174–91 (Pt. 1), 303–18 (Pt. 2).

Salvin, O., 1876. 'On the Avifauna of the Galapagos Archipelago' in *Trans. Zool.Soc.Lond.* IX:447–510.

Sato, A., Colm, O. U., Figueroa, F., Grant, P. R., Grant, B. Rosemary, Tichy, H. and Klein, J., 1999. 'Phylogeny of Darwin's Finches as revealed by mtDNA sequences' in *Proc.Nat.Acad. Sci.USA* 96:5101–6.

Sclater, P. L., 1865. 'Notes on Kruper's Nuthatch, and on the other known species of the genus *Sitta*' *in Ibis*, Series 2, 1:306–11.

Sclater, P. L., 1871. 'Remarks on the Avifauna of the Sandwich Islands' in *Ibis*, Series 3, 1:356–62.

Sclater, P. L. ed., 1890. 'Review of Tristram's Catalogue of his Collection' in *Ibis*, Series 6, 2:121.

Sclater, P. L., 1909. 'Canon H. B. Tristram. Biographical notices of the original members of the British Ornithologists' Union' in *Ibis*, Series 9, 2:153–6.

Seebohm, H., 1890. *The Birds of the Japanese Empire*. (R. H. Porter).

Seebohm, H., 1896. *Coloured Figures of the Eggs of British Birds*. (Pawson & Brailsford).

Select Committee, 1873. *Report on Wild Birds Protection*. (House of Commons).

Sharpe, R. B., 1872. 'Descriptions of some New Species of Birds in the National Collection' in *Ann.Nat.Hist.*, Series 4, 10:450.

Sharpe, R. B., 1879. *Catalogue of the Passeriformes or Perching Birds in the collection of the British Museum*. IV: Cichlomorphae: Part 1:232–3. (British Museum).

Sharpe, R. B., 1890. *Catalogue of the Birds of the British Museum*. XIII: Sturniformes 582. (British Museum).

Sharpe, R. B., 1901. *A Handlist of the Genera and Species of Birds*. III:232. (British Museum).

Sharpe, R. B., 1906. *The History of the Collections contained in the Natural History Departments of the British Museum*. Birds 2 (3):79–515. (British Museum).

Sharpe, R. B., 1907. 'Presidential Address' in *Proc.4th Int.Orn.Cong.*, 90–143. (Dulau & Co.).

Shepherd, L. D. and Lambert, D. M., 2006. 'Ancient DNA and Conservation: lessons from the endangered Kiwi of New Zealand' in *Molecular Ecology*, 17:2174–84.

Shirihai, H., 1996. *The Birds of Israel*. (Academic Press).

Sloane Collection, *MS 750. A History of the Bermudas from its first settlement to 1623*.

Smith, J., 1624. *The Generall Historie of Virginia, New England and the Summer Isles*. (James Machose & Sons).

Spurling, H., 1995. *The Life of I. Compton-Burnett*. (Richard Cohen Books).

Temperley, G. W., 1951. 'A History of the Birds of Durham' in *Trans.Nat. Hist.Soc. Northumberland and Durham*. IX:3–296.

Tristram, H. B., 1853a. 'Occurrence of the Little Auk (Uria Alle) in the City of Durham' in *Zoologist* 11:3755.

Tristram, H. B., 1853b. 'Occurrence of the Jacamar (*Galbula ruficauda*) in Lincolnshire' in *Zoologist* 11:3907.

Tristram, H. B., 1853c. 'Occurrence of the Great Snipe (*Scolopax major*) in Durham' in *Zoologist* 11:3911.

Tristram, H. B., 1859a. 'On the Birds of Southern Palestine' in *Ibis*, Series 1, 1:22–49.

Tristram, H. B., 1859b. 'Characters of apparently New Species of Birds collected in the Great Desert of the Sahara, southward of Algeria and Tunis' in *Ibis*, Series 1, 1:57–9.

Tristram,.H. B., 1859c. 'On the Ornithology of North Africa' in *Ibis*, Series 1, 1:153–62 (Pt. 1), 277–301 (Pt. 2), 415–35 (Pt. 3a.).

Tristram, H. B., 1860a. 'On the Ornithology of North Africa' in *Ibis*, Series 1, 2:68–83 (Pt.3b), 149–65 (Pt.4).

Tristram, H. B., 1860b. *The Great Sahara*. (J. Murray).

Tristram, H. B., 1860c. *Presidential Address to the Members of the Tyneside Naturalists' Field Club*, (Newcastle).

Tristram, H. B., 1862. 'On the Ornithology of Palestine' in *Ibis*, Series 1, 4:277–9.

Tristram, H. B., 1863. 'Pallas's Sand-Grouse in Durham' in *Zoologist* 21:86–9.

Tristram, H. B., 1865a. 'Exhibition of New Birds of Palestine' in *Proc.Zool. Soc.Lond* (Scientific Meetings), 169–70.

Tristram, H. B., 1865b. 'The Dead Sea' in *Zoologist* 23:9705.

Tristram, H. B., 1865c. *The Land of Israel. A Journal of Travels in Palestine.* (S.P.C.K.).

Tristram, H. B., 1865d. 'On the Ornithology of Palestine' in *Ibis*, Series 2, 1:67–83, 241–63.

Tristram, H. B., 1866. 'On the Ornithology of Palestine' in *Ibis*, Series 2, 2:59–88, 280–92.

Tristram, H. B., 1867a. 'On the Ornithology of Palestine' in *Ibis*, Series 2, 3:73–97, 360–71.

Tristram, H. B., 1867b. *The Natural History of the Bible.* (S.P.C.K.).

Tristram, H. B., 1868a. 'On the Ornithology of Palestine' in *Ibis*, Series 2, 4:204–15, 321–35.

Tristram, H. B., 1868b. 'The Zoological Aspects of the Grouse Disease' in *British Association Report*, 97. (Dundee 1867, J. Murray).

Tristram, H. B., 1871. *The Topography of the Holy Land.* (S.P.C.K.).

Tristram, H. B., 1873. *The Land of Moab.* (J. Murray).

Tristram, H. B., 1877. 'The Story of the Isles of the Sea, told by the Fowls of the Air' in *Good Words*, XVIII: 67–72, 209–13, 226–9, 337–41, 395–400, 560–8, 606–8, 706–9.

Tristram, H. B., 1882. 'Ornithological notes on a journey through Syria, Mesopotamia and Southern Armenia' in *Ibis*, Series 4, 6:402–419.

Tristram, H. B., 1884. *The Fauna and Flora of Palestine.* (Committee of the Palestine Exploration Fund).

Tristram, H. B., 1889. *Catalogue of a Collection of Birds belonging to H. B. Tristram, D.D.,LL.D.,F.R.S.* (Durham Advertiser).

Tristram, H. B., 1893. 'Differentiation, Migration and Mimicry' in *Zoologist* 51:316–86. (Presidential Address to the Biological Section of the British Association, Nottingham, September).

Tristram, H. B., 1895. *Rambles in Japan.* (The Religious Tract Society).

Tristram, H. B., 1902a. 'Letter on the Cahow of the Bermudas' in *Ibis*, Series 8, 2:525.

Tristram, H. B., 1902b. 'The "cahowe" of the Bermudas' in *Ann.& Mag. Nat.Hist.*(7) 9:447–8.

Tristram, H. B., 1905. 'Birds' in Page, W., *The Victoria History of County Durham*, 1:175–91.

Tristram, H. B., 1906. *Tawny Owls on the Banks. An Appeal.* (Durham County Advertiser, 16 February).

Tristram, Louisa Hely-Hutchinson, 1898. 'Recollections'. (Unpublished MS. Durham University Library).

Verrill, A. E., 1901. 'The Story of the Cahow. The Mysterious Extinct Bird of the Bermudas' in *Popular Science Monthly* 60:23–30.

Wagstaffe, R., 1978. *Type Specimens of Birds in the Merseyside County Museums.* (Merseyside County Museums).

Whellan, W., 1855. *History, Topography and Directory of Northumberland.* (Whittaker & Co.).

Whitaker, J. I. S., 1905. *The Birds of Tunisia.* (R. H. Porter).

Williams, K. F., 2010. 'Scarce Bird Book of 1848' in *Bull.Jourdain Soc.* 16:226–7.

Wilson, Alexander, 1808 to 1853. (var. edns). *American Ornithology.* (Bradford & Inskeep)..

Wingate, D. B., 1982. 'Successful re-introduction of the Yellow-crowned Night Heron as a nesting resident on Bermuda' in *Colonial Waterbirds* 5:104–15.

Witherby, H. F., Jourdain, F. C. R., Ticehurst, N. F. and Tucker, B. W., 1938–41. *The Handbook of British Birds.* (H. F. & G. Witherby).

Wollaston, A. F. R., 1921. *Life of Alfred Newton.* (John Murray).

Yarrell, W., 1871–85. *History of British Birds*, 4th edn. 3:201. (Van Voorst).

# Appendix 1: Tristram's Life Table

| Year | Residence | Event |
|------|-----------|-------|
| 1822 | Eglingham | Born 11 May in Eglingham, Northumberland. |
| 1830 | Eglingham | Tristram's mother died in childbirth. |
| 1932 | Eglingham | On Tour—Durham, York, Grantham, London, Weymouth, Bath, Bristol. |
| 1836 | Eglingham | So far Henry had been tutored by his father and his father's curate. |
| 1837 | Durham | Tristram's father died. Entered Durham School. |
| 1840 | Durham | Entered Oxford (Lincoln College) in April. |
| 1844 | Durham | B.A., graduated in Classics. Tutored Charles Harford in Switzerland & Italy. |
| 1845 | Devon | Ordained Deacon and became Curate at Morchard Bishop. |
| 1846 | Devon | M.A. Awarded. Visited France. Left for, and appointed Chaplain of, Bermuda |
| 1849 | Bermuda | Left Bermuda in May. Returned to England via Canada and North America. Appointed Rector of Castle Eden. |
| 1850 | Castle Eden | Married Eleanor Mary Bowlby, 5 February Visited Orkney. Mary Gertrude Tristram, born 28 November 1850. |
| 1852 | Castle Eden | Visited Norway, Finnmark and Denmark. Louisa Hely-Hutchinson, born 25 April 1852. |
| 1853 | Castle Eden | Tristram's first publications in the *Zoologist* and gets a Great Auk egg. Eleanor Frances (Ella) Tristram, born 20 December 1853. |
| 1855 | Castle Eden | Ill again—went to Torquay with his step-mother. Winter in Algeria (1). Charlotte Eliza, born 12 May 1855. |
| 1857 | Castle Eden | Left North Africa for England 11 July. Winter Cruise begun (Eastern Mediterranean). |
| 1858 | Castle Eden | Winter cruise to the Mediterranean including Turkey and Palestine (1). Katherine Alice Salvin (Katie), born 29 April 1858. |
| 1859 | Castle Eden | BOU and *Ibis* founded. *Ibis* papers. Darwin supported in print. |

# APPENDIX 1

| Year | Residence | Event |
|---|---|---|
| 1860 | Castle Eden | President, Tyneside Naturalists' Club. Oxford 'Debate'. *Great Sahara* published. Christiana (Chrissie), born 11 February 1860. Appointed Rector of Greatham and Master of Greatham Hospital. |
| 1861 | Greatham | Dublin; shot a Rock Pipit. Henry Barrington (Harry or later Tim), born 5 September 1861. |
| 1863 | Greatham | Over-winter expedition to Palestine (2). |
| 1864 | Greatham | Torquay; shot a Pratincole. Frances Anne Georgina (Annie), born 31 May 1864. |
| 1865 | Greatham | *The Land of Israel* published. J. H. Gurney Jr arrived as pupil for 2-3 years. |
| 1867 | Greatham | *Natural History of the Bible* published. France (Boulogne), shot Pochard. Wolverhampton. |
| 1868 | Greatham | First communications between Darwin and Tristram. Elected F.R.S. |
| 1870 | Greatham | Appointed Canon of Durham Cathedral. |
| 1871 | Greatham | *The Topography of the Holy Land* published. |
| 1872 | Greatham | Departure of expedition to Palestine (3), 10 January. |
| 1873 | Durham City | *The Land of Moab* published. Appointed Residentiary Canon. |
| 1877 | Durham City | Riga, Moscow and Nijni Novgorod. |
| 1881 | Durham City | Egypt, Palestine (4), Lebanon, Mesopotamia and Armenia. |
| 1884 | Durham City | Appointed Masonic Grand Chaplain for England. |
| 1885 | Durham City | Appointed Masonic Deputy Provincial Grand Master for Durham. |
| 1888 | Durham City | Canary Islands. |
| 1889 | Durham City | Palma, Madeira and the Canary Islands. |
| 1890 | Durham City | Canary Islands. |
| 1891 | Durham City | Ismailia, Red Sea, Ceylon, Singapore, Hong Kong, Shanghai to Japan. |
| 1893 | Durham City | Presidential Address to Section D of the British Association. |
| 1894 | Durham City | Palestine (5). |
| 1895 | Durham City | *Rambles in Japan* published. With youngest daughter in the Tyrol. |

| Year | Residence | Event |
|---|---|---|
| 1896 | Durham City | Sale of the Skin Collection to Liverpool Museum. |
| 1897 | Durham City | In Greece and Turkey with two daughters. Palestine (6). Leg broken. |
| 1899 | Durham City | Canary Islands |
| 1901 | Durham City | Canon and Mrs Tristram celebrated their Golden Wedding, 5 February. |
| 1903 | Durham City | Tristram attended Anniversary Meeting of the BOU in May in London. |
| 1904 | Durham City | Sicily with Annie. |
| 1905 | Durham City | Tristram attended the Cambridge Meeting of the British Association. |
| 1906 | Durham City | Tristram died on 8 March. |

# Appendix 2: Correspondence associated with Tristram

Sources of letters in the table below:

| | |
|---|---|
| **9839/1T** | Cambridge University Library, Newton correspondence |
| **DCP** | Darwin Correspondence Project, Cambridge University Library |
| **WCP** | Wallace Correspondence Project, Cambridge University Library |
| **NMS** | National Museum of Scotland, Edinburgh |
| **NEWHM** | Newcastle Hancock Museum |
| **Tristram Family Private Collections** | |
| **Chilton 2014** | In: Chilton, Glen, 2014. *The Curse of the Labrador Duck* |
| **Crieve 1885** | In: Grieve, S., 1885. *The Great Auk or Garefowl* |
| **Rowley 1875–7** | In: Rowley, G.D., 1875–7. *Ornithological Miscellany* |
| **Wollaston 1921** | In: Wollaston, A.F.R., 1921. *The Life of Alfred Newton* |

| Source | From | To | Date | Content |
|---|---|---|---|---|
| 9839/1T/199 | A. Tristram | A. Newton | 21 Sep 1889 | Reply to request for loan of skins. |
| Tristram Family | J. Hancock | H. B. T | 26 Nov 50 | Sending specimens of Ural Owl. Eggs of Sandgrouse. |
| Wollaston 1921 | A. Newton | H. B. T. | 1 Sep 1857 | Glories of African Generals of all times sink to nothing when compared with yours. |
| 9839/1T/200 | E. M. Tristram | A. Newton | 6 Nov 1857 | Loan of eggs from Algerian collection. |
| 9839/1T/201 | E. M. Tristram | E. Newton | 5 Mar 1872 | Reply to request for loan of skins. |
| 9839/T/202 | H. B. T. | J. Wolley | 4 Sep 1854 | Recommending Drummond as exploration companion. |
| 9839/1T/203 | H. B. T. | A. Newton | 15 Jan 1855 | King Duck. Catalogue for Wolley's Sale. |
| 9839/1T/204 | H. B. T. | A. Newton | 9 May 1855 | Cirl Bunting, Black Redstart and Ducks. Sent from Torquay. |
| 9839/1T205 | H. B. T. | E. Newton | 20 Oct 1857 | Mrs Tristram will send eggs. Delay in sailing and sent from Bristol. |
| Wollaston 1921 | A. Newton | E. Newton | 2 May 1858 | Progress in Iceland. |
| Wollaston 1921 | A. Newton | E. Newton | 28 May 1858 | Wolley is more sanguine about our success than I am (about finding the Great Auk). |
| 9839/1T/206 | J. Wolley | H. B. T. | 23 Jun 1858 | Accepting Mrs Tristram's invitation on return from Iceland. Sent from Iceland. |
| Wollaston 1921 | A. Newton | E. Newton | 16 Aug 1858 | The result in short was nothing (in not finding the Great Auk). |
| Wollaston 1921 | A. Newton | H. B. T. | 24 Aug 58 | Informing him of the Darwin/Wallace paper. For your paper you must first consult it. |
| Wollaston 1921 | A. Newton | H. B. T. | 10 Dec 1858 | Nearly lost Wolley over name of 'Ibis'. Things look promising for the journal. |

APPENDIX 2 289

| Source | From | To | Date | Content |
|---|---|---|---|---|
| Private Collection | J. Wolley | H. B. T. | 11 Apr 1859 | Details of egg clutches apparently previously sent to Tristram. |
| Wollaston 1921 | A. Newton | A. C. Smith | 18 Jul 1859 | Messrs. Salvin and Newton vary so in their accounts (of Palestine). |
| 9839/1T/207 | A. Newton | H. B. T. | 9 Mar 1860 | Publication of Darwin's book. Tristram's Tyneside Meeting address. |
| 9839/1T/208 | A. Newton | H. B. T. | 9 Apr 1860 | Thanking for newspaper cuttings re Tyneside address. Fair comments on Darwin. |
| Wollaston 1921 | A. Newton | H. B. T. | 25 Apr 1860 | I have not got a Magpie's egg with which to match (it)—the egg of a Nutcracker. |
| DCP2852 | J. D. Hooker | C. Darwin | Jul 1860 | Speech at the Oxford Debate. |
| Tristram Family | G. D. Rowley | H. B. T. | 23 Jul 1860 | Thanks for copy of the address to Tyneside Naturalists' Society & discussion of contents. |
| Wollaston 1921 | A. Newton | E. Newton | 25 Jul 1860 | Tristram waxed exceeding wrath . . . and declared himself more and more anti-Darwinian. |
| 9839/1T/209 | A. Newton | H. B. T.* | 30 Jul 1860 | Newton now 'Darwinoid'. Tristram's position compared with Chancery suit. |
| Wollaston 1921 | A. Newton | H. B. T. | 31 Jul 1860 | You are not fit to go to Germany without a Keeper. This renovation of Lamarck . . . |
| NMS | H. E. T. | J. A. Harvie-Brown | Jul 186? | Adverse effects of publications on bird protection. |
| Wollaston 1921 | A. Newton | H. B. T. | 18 Aug 1860 | Newton's acquisition of the egg of a Great Auk. |

| Source | From | To | Date | Content |
|---|---|---|---|---|
| Clerk Uni. HP 15:117 | T. H. Huxley | Frederick Dyster | 9 Sep 1860 | I unhesitatingly affirm(ed) my preference for the ape … there was unextinguishable laughter |
| 9839/1T/? | A. Newton | H. B. T. | 30 Nov 1860 | Bree's book on Darwinism. |
| In Grieve 1885 | W. Proctor | R. Champley | 28 Feb 1861 | The acquisition of a mounted Great Auk by Durham Museum. |
| Wollaston 1921 | A. Newton | H. B. T. | 22 Aug 1861 | Newton's visit to Germany. Sent from Germany. |
| Private Collection | G. D. Rowley | H. B. T. | 3 Dec 1861 | Exchange of Eggs. Purchased 3 Nutcracker's eggs at Wheelwright's sale for £3–9–0d. |
| Wollaston 1921 | A. Newton | E. Newton | 25 Dec 1861 | Newton's discovery of ten eggs of the Great Auk. |
| Wollaston 1921 | A. Newton | H. B. T. | 3 Jul 1862 | Bewick is published … write us a notice of it for the Ibis. |
| Wollaston 1921 | A. Newton | H. B. T. | 11 Jul 1863 | From Lapland complaining about packing eggs. Reference to Tristram and Canterbury. |
| Wollaston 1921 | A. Newton | E. Newton | 25 Mar 1864 | 20 new spp. of birds found in Palestine—including Grackle (Letter Tristram to Salvin). |
| 9839/1T/210 | H. B. T. | A. Newton | 22 Sep 1864 | I congratulate you myself, the *Ibis* and the world on your safe return. |
| 9839/1T/211 | H. B. T. | A. Newton | 11 Oct 1864 | I cannot think of making pecuniary decisions with an old friend about eggs. |
| 9839/1T/212 | H. B. T. | A. Newton | 3 Feb | Last date for delivery of *Ibis* paper 14 February. |
| 9838/1T/213 | H. B. T. | A. Newton | 10 Feb | I will if I can or rather would if I could. |
| 9839/1T/215 | H. B. T. | A. Newton | 17 Oct 1865 | Greek Partridges. 100 of 300 with spots. Lost baggage. Palestine parcels valued at £200. |

APPENDIX 2

| Source | From | To | Date | Content |
|---|---|---|---|---|
| Wollaston 1921 | A. Newton | H. B. T. | 1 Mar 1866 | By this time tomorrow I shall be Man or Mouse. Election to the Chair in Cambridge. |
| 9839/1T/216 | H. B. T. | A. Newton | 21 May 1866 | I wish I could get away again this summer. I am on my way back from Birmingham. |
| 9839/1T/217 | H. 3. T. | A. Newton | 17 Aug 1866 | Hope to be at Nottingham 8.25pm. Tuesday. |
| 9839/1T/218 | H. B. T. | A. Newton | 18 Sep 1866 | Waxwing eggs bought by Buckley in Montreal at 1/- apiece. |
| 9839/1T/219 | H. B. T. | A. Newton | 5 Oct 1866 | Hope to send paper by Nov. 15th. You cannot tell how I am never off my legs. |
| 9839/1T/220 | H. B. T. | A. Newton | 20 Oct 1866 | Blyth on Jerdon's 'Birds of India'. |
| 9839/1T/222 | H. B. T. | A. Newton | 12 Nov 1866 | Wish I could get at you but will be in Birmingham only 48hrs. |
| 9839/1T/221 | H. B. T. | A. Newton | 18 Nov 1866 | Visit of the Gurneys. Sent MS for *Ibis*. Can't get rid of my cough! |
| 9839/1T/223 | H. B. T. | A. Newton | 14 Nov 1866 | Gurney ill; Mrs T. Looking after him. I do not know when I shall get S. P. C. K. Fine. |
| 9839/1T/224 | H. B. T. | A. Newton | 22 Nov 1866 | Cryptic note relating to Gould. |
| 9839/1T/225 | H. B. T. | A. Newton | 29 Nov 1866 | Gurney better. Wish for publication of list of Birds of Europe to supercede *Zoologist*. |
| 9838/1T/226 | H. B. T. | A. Newton | 21 May 1867 | I will try to send up paper for this *Ibis*. Send PO for Constantinople. |
| 9839/1T/227 | H. E. T. | A. Newton | 9 Oct 1867 | Paper supporting Newton's Darwinism. |
| 9839/1T/228 | H. E. T. | A. Newton | 4 Nov (1867) | Off next week to speak and preach in Brighton. |
| 9839/1T/229 | H. E. T. | A. Newton | 2 Mar 1868 | Complaining of church politics. |

| Source | From | To | Date | Content |
|---|---|---|---|---|
| DCP? | C. Darwin | H. B. T. | 4 Jun 1868 | Request for information concerning desert birds. Newly discovered letter. |
| DCP6234 | H. B. T. | C. Darwin | 6 Jun 1868 | Reply later. Thanking you for your unexpected kindness (signing FRS certificate). |
| DCP6262 | H. B. T. | C. Darwin | 1 Jul 1868 | On the coloration of 26 species of Saharan birds. |
| 9839/1T/214 | H. B. T. | A. Newton | 8 Jul 1868 | Lot 259 eggs not mine. From India? Have you seen my books? I have only one copy. |
| DCP6351 | H. B. T. | C. Darwin | 5 Sep 1868 | Sexual difference in plumage of birds; various species compared. |
| 9839/1T/230 | H. B. T. | A. Newton | No date | I have just sent off this incorrect squib o' Billy Gladston's to an Irish friend . . . |
| 9839/1T/231 | H. B. T. | A. Newton | 2 Sep (1870) | I had a tremendously long Deutch letter from Baldanus . . . does he read English? |
| 9839/1T/232 | H. B. T. | A. Newton | 1 Jan 1871 | Palestine Exploration Fund business. |
| NMS | H. B. T. | J. A. Harvie-Brown | 10 Feb 1871 | Advice on visiting Norway and on the nesting of the Green Sandpiper. |
| 0839/1T/233 | H. B. T. | A. Newton | 15 Feb 1871 | If PEF pay my expenses I will not object to an extra couple of days with you. |
| 9839/1T/234 | H. B. T. | A. Newton | 20 Feb 1871 | Meeting arrangements. |
| 9839/1T/235 | H. B. T. | A. Newton | 23 Feb | Meeting arrangements. |
| 9839/1T/236 | H. B. T. | A. Newton | 10 Mar 1871 | Can you send me types of Strickland's *Phyllopneuste brevirostris*? |

| Source | From | To | Date | Content |
|---|---|---|---|---|
| 9839/1T/237 | H. B. T. | A. Newton | 28 Mar 1871 | I return *P. brevirostris* with many thanks. I find I have five exactly like them. |
| NEWHM:1967. H67. | W. E. Brooks | J. Hancock | 12 Aug 1871 | Request to ask Tristram about 'getting Hewitson to do a plate' of Indian eggs for *Ibis*. |
| 9839/1T/238 | H. B. T. | A. Newton | 11 Jul 1872 | *Didunculus* and *Nestor*. |
| Wollaston 1921 | A. Newton | H. B. T. | 29 Jul 1872 | Wildfowl Bill, followed . . . by one for the regulation of birdcatchers would be better. |
| Tristram Family | British Museum | H. B. T. | 31 Oct 1872 | "Enclosed are the two lithographs of the new Moabite . . . " |
| Tristram Family | Lord Lilford | H. B. T. | 24 Dec 1872 | I have a pair of *Struthides cinerea* alive. Nest of *Ph. superciliosus*. |
| Newcastle 0533 | A. Newton | J. Hancock | 18 Feb 1873 | Letter from Tristram on Sandgrouse in the UK |
| Newcastle 0544 | A. Newton | J. Hancock | 8 May 1873 | Tristram tells me of a specimen (Firecrest) some years ago at Brancpeth. |
| NMS | H. B. T. | J. A. Harvie-Brown | 7 Jan 1873 | Eggs of Terek Sandpiper and Ptarmigan. Sale of eggs of Yellow-browed Warbler. |
| 9839/1T/239 | H. B. T. | A. Newton | Copy 1873 | I am disappointed and surprised at this stupidity of the Bird Committee. |
| 9839/1T/240 | H. B. T. | A. Newton | 9 Aug 1873 | *Tinnunculus punctatus* and *Nectarinia*. |
| Tristram Family | A. Newton | H. B. T. | 11 Aug 1673 | "Red Lions to roar". Landbirds of the Seychelles. Col. Pike has just written a stupid book. |

| Source | From | To | Date | Content |
|---|---|---|---|---|
| NEWHM:1967.H67. | W. E. Brooks | J. Hancock | 26 Aug 1873 | Exchange loans of skins with Tristram who is returning dark brown 'Aquila naevoides'. |
| 9839/1T/241 | H. B. T. | A. Newton | 15 Oct (1873) | I have got a very fair skin of my *Didunculus*. |
| 9839/1T/242 | H. B. T. | A. Newton | 18 Oct 1873 | Birds of the Celebes and Sumatra. |
| 9839/1T/243 | H. B. T. | A. Newton | 22 Nov 1873 | The Bishop has appointed me today to a Canon Residentiary. |
| WCP5254 | A. R. Wallace | H. B. T. | 13 Jan 1874 | I am trying at a book on the Geog. Distribution of Animals. |
| DCP9295 | C. Darwin | H. B. T. | 16 Feb 1874 | Asking H. B. T. to sign Royal Society certificate for proposal of Robert Swinhoe's Fellowship. |
| DCP9371a | C. Darwin | H. B. T. | 26 Feb (1874) | Asking about certificate for Robert Swinhoe. |
| DCP9330 | C. Darwin | H. B. T. | 3 Mar (1874) | Thanking H. B. T. for signing Robert Swinhoe's certificate. |
| 9389/1T/244 | H. B. T. | A. Newton | 12 Mar 1874 | Parliamentary duties. In London 24 April for a fortnight. |
| 9839/1T/245 | E. M. Tristram | A. Newton | 3 Jul (1874) | Writing to say H. B. T. cannot 'notch' a pen—attack of bronchitis. |
| Tristram Family | A. R. Wallace | H. B. T. | 11 Oct 1874 | Do you know of any books on migration? Queries on classification. |
| 9839/1T/246 | H. B. T. | A. Newton | 15 Aug 1874 | Candidates for the chair of Biology in Newcastle. |
| 9839/1T/247 | H. B. T. | A. Newton | 20 Aug 1874 | I went to Newcastle on Monday for the election of the Professor. |
| NEWHM:1967.H67. | W. E. Brooks | J. Hancock | 26 Aug 1874 | Dark brown 'Aquila naevoides' not yet returned; if you have it send it direct. |
| 9839/1T/248 | H. B. T. | A. Newton | 29 Aug 1874 | I am by no means satisfied by the results of yesterday. |
| 9839/1T/249 | H. B. T. | A. Newton | 3 Sep 1874 | I am hopelessly in arrears with my work. |

# APPENDIX 2

| Source | From | To | Date | Content |
|---|---|---|---|---|
| 9839/1T/250 | H. 3. T. | A. Newton | 28 Oct 1874 | Is it possible Edward (Newton) has been in England for a month & I have not heard? |
| NEWHM:1967.H67. | W. E. Brooks | J. Hancock | 12 Mar 1875 | Much on variation in Yellow Wagtails. More Eagles to Tristram (from India). |
| NEWHM:1967.H67. | J. Hancock | W. E. Brooks | 8 Apr 1875 | There is too great a love for making species—'Race principle' for Yellow Wagtails. |
| NMS | H. B. T. | J. A. Harvie-Brown | 30 Sep 1875 | Congratulations on the success of the expedition to the Petchora with Seebohm. |
| Private Collection | G. D. Rowley | H. B. T. | 16 Dec 1875 | Loan of book. Request for article on desert larks for Rowley's Miscellany. |
| NMS | H. B. T. | J. A. Harvie-Brown | 21 Feb 1876 | Thank you for two interesting papers. Possible exchange of eggs. |
| 9389/1T/251 | H. B. T. | A. Newton | 7 May 1876 | Notes on Birds of Palestine and intent on visiting Glasgow (for meeting). |
| In Rowley 1877 | D. G. Elliot | D. G. Rowley | 31 May 1876 | Capture of Labrador Duck. |
| 9389/1T/252 | H. B. T. | A. Newton | 19 Jun 1876 | I shall send you a paper on the bird life of the South Pacific. |
| NMS | H. B. T. | J. A. Harvie-Brown | 19 Jun 1876 | Mislaying bill for eggs from Small. Sending reprints. |
| 9389/1T253 | H. E. T. | A. Newton | 26 Jul 1876 | I must talk about the eggs . . . but you must come to Durham. Sent from Southampton. |

| Source | From | To | Date | Content |
|---|---|---|---|---|
| NEWHM:1967. H67. | J. Hancock | Species note | Probably 1876 | This new species is the same as *Aquila naevioides* and *A. flavescens*. |
| NEWHM:1996. H74.4 | H. Seebohm | J. Hancock | 29 Jan 1877 | Casts of Great Auks eggs sent to Seebohm. An interest from Tristram indicated. |
| NEWHM:1967. H67. | H. B. T. | J. Hancock | 3 Feb 1877 | Request for suggestions for a successor to Proctor in Durham Museum. |
| NEWHM:1967. H67. | J. Hancock | H. B. T. | 1877 | Cannot suggest anyone who would take Proctor's post at £30 a year—suggest £60. |
| NMS | H. B. T. | J. A. Harvie-Brown | 15 May 1877 | Synonymy in the Genus *Hippolais*. |
| NMS | H. B. T. | J. A. Harvie-Brown | 18 May 1877 | Exchange of New Zealand specimens and papers. |
| NEWHM:1967. H67. | G. R. Hall | J. Hancock | 25 Jun 1877 | Tristram has given son George (at Durham School) some eggs. Do you have any spare? |
| NMS | H. B. T. | J. A. Harvie-Brown | 22 Oct 1877 | Thanks for paper. Have you seen Seebohm yet? |
| 9389/1T/254 | H. B. T. | A. Newton | 27 Jan 1879 | I have skeleton of Kaga. Do you want it? Layard has sent a lot of specimens. |
| 9389/1T/255 | H. B. T. | A. Newton | 28 Jan 1879 | I found I have made a mistake. |
| NEWHM:1967. H67. | H. B. Hewitson | J. Hancock | 7 Sep 1879 | Illustrated catalogue of the Great Auk—permission from owners of all Great Auk eggs. |

# APPENDIX 2

| Source | From | To | Date | Content |
|---|---|---|---|---|
| NEWHM:1967. H67. | J. Hancock | H. B. Hewitson | 19 Sep 1879 | Come and draw the Great Auk's egg when you visit Tristram. |
| 9389/1T/256 | H. B. T. | A. Newton | 1 Apr 1882 | I am quite free from Whitsun Monday up to 19 July so I can go whenever you like. |
| NEWHN:1967. H67. | W. E. Brooks | J. Hancock | 30 Jun 1882 | Tristram has sold all his eggs. My collection is in Johnson's hands in Clavering Place. |
| Wollaston 1921 | A. Newton | A. C. Smith | 5 Oct 1882 | Woodall agreed to take Tristram and myself for a cruise. Newton injured his good leg. |
| Private Collection | Heinrich Gatke | H. B. T. | 15 Jul 1884 | Thanks for this beautiful work [probably *Fauna and Flora of Palestine*]. Wheatears. |
| 9389/1T/257 | H. B. T. | A. Newton | 28 Feb 1885 | You have sent me a valuable addition to my bird lore in your article on Ornithology. |
| NEWHM:2002. H1047.9 | H. B. T. | W. E. Brooks | 20 Dec 1885 | Sending Broad-billed Sandpiper skin and copy of Egg-Catalogue. Note on Labrador Duck. |
| Tristram Family | Palestine Ex. Fund | H. B. T. | 17 Feb 1886 | I am pleased you like my prospectus. |
| Tristram Family | Sir J. Kirk? | H. B. T. | 2 Nov 1887 | Museum collections. |
| Wollaston 1921 | A. Newton | H. B. T. | 2 Jan 1888 | Original founders of B. O. U. Newton's belief that *Ibis* might have 'its youth renewed'. |
| Wollaston 1921 | A. Newton | H. B. T. | 2 Feb 1888 | Let me have any letters showing my frame of mind before publication of the 'Origin'. |

| Source | From | To | Date | Content |
|---|---|---|---|---|
| Wollaston 1921 | A. Newton | H. B. T. | 26 Nov 1888 | The self-animation of our biologists . . . is beyond any praise I can bestow. |
| Private Collection | H. B. T. | A. Newton | 29 Jan 1889 | Broad-billed Sandpiper . . . they say one of the young Rothschilds is taking up ornithology. |
| NEWHNM:2002. H1047. | H. B. T. | W. E. Brooks | 29 Jan 1889 | Mention of Wolley, Sclater, Hancock and Seebohm. Note by A. C. Brooks re Labrador Duck. |
| 9389/1T/258 | H. B. T. | A. Newton | 15 Feb 1889 | I may possibly go to the Cape Verde Islands next week. |
| Wollaston 1921 | A. Newton | H. B. T. | 16 Feb 1889 | I don't care a button whether you declare these islands to be Palaearctic or Ethiopean. |
| Wollaston 1921 | A. Newton | H. B. T. | 29 Dec 1890 | It really is a pity you can't come home via the Sandwich Islands [from Japan]. |
| Private Collection | D. Jamal | H. B. T. | 19 Aug 1891 | Tristram's Dragoman. Resigned Mr Cook's emplyment; request for recommendations. |
| 9389/1T/259 | H. B. T. | A. Newton | 5 Jun 1892 | I leave town for Brindisi on Monday morning at 7.0am. |
| Tristram Family | ? | H. B. T. | 14 Jun 1892 | Song of the Yellow Hammer. |
| Tristram Family | Lord Rothschild | H. B. T. | 12 Jul 1892 | Mr Dresser tells me that you identified some of my Sandwich Island birds. |
| Wollaston 1921 | A. Newton | H. B. T. | Sep 1892 | I think it can hardly be inconsistent with Christian principles to hate bores! |
| 9389/1T/260 | H. B. T. | A. Newton | No date | Lost luggage. |

| Source | From | To | Date | Content |
|---|---|---|---|---|
| Wollaston 1921 | A. Newton | H. B. T. | 23 Oct 1893 | I admire the judgement of your doctor in prohibiting the two 'p's'—preaching & politics. |
| NMS | H. B. T. | J. A. Harvie-Brown | 11 Jan 1896 | Recommending "young? Knight", pupil of Cullingford of Durham. Death of Drummond-Hay. |
| Private Collection | A. Newton | H. B. T. | 6 Feb 1896 | H. B. T. not replied to letters. Mention of future visit of H. B. T. to Canary Islands. |
| Tristram Family | Picton? | H. B. T. | 1 Oct 1896 | Sale of New Zealand birds. |
| Wollaston 1921 | A. Newton | H. B. T. | 20 May 1897 | Edward Newton was an excellent bird-nester. |
| In Chilton, Glen 2014 | H. B. T. | ? | 1901 | I never had much confidence in the authenticity of the Labrador Duck. |
| BM, Tring | H. B. T. | T. Parkin | 4 Apr 1903 | I now enclose my oological biography. [Now in Tristram's Egg Catalogue]. |
| Wollaston 1921 | A. Newton | H. W. Fielden | 31 Oct 1905 | Tristram introduced as the Emperor Louis Napoleon to the Mayor of Brighton. |
| Wollaston 1921 | H. B. T. | A. Newton | 25 Feb 1906 | Last letter to Newton. |
| Wollaston 1921 | A. Newton | H. B. T. | 26 Feb 1906 | If the letter I have just received is really to be the *last* . . . |

# Appendix 3: Tristram's Ornithological Publications

**1853**

Occurrence of the Little Auk (*Uria alle*) in the City of Durham, *Zoologist* 11:3755

Occurrence of the Jacamar (*Galbula ruficauda*) in Lincolnshire, *Zoologist* 11:3907

Occurrence of the Great Snipe (*Scolopax major*) near Durham, *Zoologist* 11:3911

Notes on the Grasshopper Warbler, *Zoologist* 11:4123

**1854**

Occurrence of the Little Olivaceous Gallinule (*Ortygometra pusilla*) at Belbriggen (Ireland), *Zoologist* 12:4295

A Little Ochraceous Gallinule at Belbriggen, *Zoologist* 12:4298–99

**1856**

A Pelican found dead on the Coast of Durham, *Zoologist* 14:5391

**1858**

Catalogue of Eggs collected in Algeria, sold at Steven's saleroom, 9 February 1858

**1859**

Occurrence of the Hobby in the Farne Islands, *Zoologist* 17:6602

On Birds Observed in Southern Palestine, *Ibis*, Series 1, 1:22–49

Characters of Apparently New Species of Birds Collected in the Sahara, *Ibis*, Series 1, 1:57–9

On the Ornithology of Northern Africa, *Ibis*, Series 1, 1:153–62 (Part 1), 277–301 (Part. 2), 415–35 (Part 3a)

**1860**

Return of a Hooded Crow to a Walled Garden in which it had been confined, *Zoologist* 18:7105

On the Ornithology of Northern Africa, *Ibis*, Series 1, 2:68–83 (Part 3b), 149–65 (Part 4)

On the Eggs of the Nutcracker and Parrot Crossbill, *Ibis*, Series 1, 2:168–71

A Few Leaves from a Naturalist's Note Book in Eastern Algeria., *Ibis*, Series 1, 2:361–75

*The Great Sahara*, London.

Presidential Address to the members of the Tyneside Naturalists' Field Club, Newcastle

**1861**

Exhibition of a Snake of Pekin & of a Series of Pipits, *Proc.Zool.Soc.Lond.* (Scientific Meetings), 391

Nesting of the Griffon Vulture in Eastern Algeria, *Zoologist* 19:7380

Ostrich Hunting in Northern Africa, *Zoologist* 19:7546–47

Letter: On Some Additions to the Birds of the Sahara, *Ibis*, Series 1, 3:414–15

**1862**

On the Ornithology of Palestine, *Ibis*, Series 1, 4:277–9

**1863**

Letter: On South European Birds, *Ibis*, Series 1, 5:364–6

Pallas' Sand Grouse in Durham, *Zoologist* 21:8689

**1864**

Exhibition of New Birds of Palestine, *Proc.Zool.Soc.Lond.* (Scientific Meetings), 169–70

Report on the Birds of Palestine, *Proc.Zool.Soc.Lond.* (Scientific Meetings), 426–56

Exhibition of a Pair of Sanderlings (*Calidris arenaria*) from Grimsey Island, Iceland, and three eggs, supposed to be those of that bird, *Proc.Zool.Soc.Lond.* (Scientific Meetings), 377

Experiences in France, *Ibis*, Series 1, 6:135–6

Letter: On the birds of Palestine, *Ibis*, Series 1, 6:229–31

**1865**

The Dead Sea, *Zoologist* 23:9705

*The Land of Israel. A Journal of Travels in Palestine*, London

On the Ornithology of Palestine, *Ibis*, Series 2, 1:67–83, 241–63

**1866**

Egyptian Geese (*Aplochen aegyptiacus*) at Stockton-on-Tees, *Zoologist* 24:25

On the Ornithology of Palestine, *Ibis*, Series 2, 2:59–88, 280–92

**1867**

On New Species of Birds from South Africa, *Proc.Zool.Soc.Lond.* (Scientific Meetings), 887

*The Natural History of the Bible*, London

On the Ornithology of Palestine, *Ibis*, Series 2, 3:73–97, 360–71

Letter: On the Pyrenian *Cinclus*, *Ibis*, Series 2, 3:466

**1868**

Letter: On *Recurvirostra sinensis*, *Ibis*, Series 2, 4:133

On the Ornithology of Palestine, *Ibis*, Series 2, 4:204–15, 321–35

On the Zoological Aspects of the Grouse Disease, Report of the British Association for the Advancement of Science, Dundee 1867, London

**1869**

On Some New South-African *Sylviidae, Ibis*, Series 2, 5:204–8
On Some African Birds, *Ibis*, Series 2, 5:434–8

**1870**

Exhibition of Skins of *Aquila naeviodes* and other Indian Eagles, *Proc.Zool. Soc.Lond.* (Scientific Meetings), 4-5
Exhibition of Skins of *Lusciola melanopogon* & other Indian warblers, *Proc. Zool.Soc.Lond.* (Scientific Meetings), 221
Letter: On the Occurrence in India of *Sylvia melanopogon, Ibis*, Series 2, 6:301
Letter: On a New Species of *Calliope, Ibis*, Series 2, 6:444
On Some Old World Species of Passerine Birds, *Ibis*, Series 2, 6:493–7

**1871**

On the Synonymy of Certain *Phylloscopi, Ibis*, Series 3, 1:109–10
On Palaearctic Passeres, *Ibis*, Series 3, 1:231–4
On Some Old World Species of Passerine birds, *Ibis*, Series 3, 1:479
*List of Birds, Gray's arrangement*. Privately printed. One folio sheet, dated 1 August 1871
*The Topography of the Holy Land*, London

**1872**

On *Syrrhaptes paradoxus* in Northumberland, *Ibis*, Series 3, 2:334–5
On a New Sylviad, *Ibis*, Series 3, 2:296–7
Letter: On *Phyllopneuste schwartzi, Ibis*, Series 3, 2:463–4
Letter: Notes on Some Passerine Birds, chiefly Palaearctic, *Ibis*, Series 3, 2:464
Notes on Sylviads. *Ann.&Mag.Nat.Hist.*, Series 4, VIII:28

## 1873

*The Land of Moab*, London
On a New Sylviad, *Ibis*, Series 3, 3:489

## 1876

Notes on a collection of Birds from the New Hebrides, *Ibis*, Series 3, 6:259–67

## 1877

The Story of the Isles of the Sea, told by the Fowls of the Air, *Good Words*, XV111:67–72, 209–13, 226–9, 337–41, 395–400, 560–8, 606–8, 706–9
List of Birds collected by the Survey Party in Palestine. Palestine Exploration Fund (PEF) Quarterly Statement, 200

## 1879

Description of a New Species of Woodpecker (*Dryocopus richardsi*) from the island of Tsu Sima, near Japan, *Proc.Zool.Soc.Lond.* (Scientific Meetings), 386–7
Notes on Collections of Birds sent from New Caledonia, from Lifu (one of the Loyalty Islands) and from the New Hebrides, *Ibis*, Series 4, 3:180–95
On a Collection of Birds from the Solomon Islands and the New Hebrides, *Ibis*, Series 4, 3:437–44
(with E. L. Layard) On a New Thrush from the Loyalty Islands, *Ibis*, Series 4, 3:471–3

## 1880

Letter: On *Asio butleri*, *Ibis*, Series 4, 4:245–6
On the Birds of the Solomon Islands, *Ibis*, Series 4, 4:246–7
Description of a New Genus and Species of Owl (*Gymnoscops insularis*) from the Seychelles, *Ibis*, Series 4, 4:456–9
Presidential Address to the Tyneside Naturalists' Field Club, Newcastle, 1-21

## 1881

Exhibition of and Remarks upon, the Skin and Eggs of a Darter, and a Skin of a Cormorant, *Proc.Zool.Soc.Lond.* (Scientific Meetings), 826–7

Description of a New Fruit-pigeon of the Genus *Carpophaga* from the Louisiade Archipelago, *Proc.Zool.Soc.Lond*, (Scientific Meetings), 996–1018

Remarks on the Notes on the Avifauna of New Caledonia and the New Hebrides by Layard and Layard, *Ibis*, Series 4, 5:132–9

Letter: On Birds from St Ambrose Island, *Ibis*, Series 4, 5:177

Notes on a Collection of Birds from the Marquesa Islands, *Ibis*, Series 4, 5:249–52

Canon Tristram's Report on the Palestine Mission, *Church Missionary Intelligenser and Record*, 6:544–5

## 1882

Description of a New Species of Land-rail *Crex sahelensis* from East Africa, *Proc.Zool. Soc.Lond.* (Scientific Meetings), 93

On the Birds of the Solomon Islands, *Ibis*, Series 4, 6:133–46

Letter: On *Erythrura cyanifrons*, *Ibis*, Series 4, 6:180–1

Ornithological Notes on a Journey through Syria, Mesopotamia and Southern Armenia, *Ibis*, Series 4, 6:402–19

Remarks concerning the Notes on the Avifauna of New Caledonia by Layard and Layard, *Ibis*, Series 4, 6:493–546

Letter: On a Halcyon from the Solomon Islands, *Ibis*, Series 4, 6:609–10

Report on the Palestine Mission. *Proceedings of the Church Missionary Society*, 83:50–9

*The Land of Israel*, 5th Revised Edition, London

## 1883

On the Position of the Acrocephaline Genus *Tartare* with Descriptions of Two New Species of the Genus *Acrocephalus*, *Ibis*, Series 5, 1:38–46

Notes on the Birds of Fanning Island, Pacific, *Ibis*, Series 5, 1:46–8

"Fauna and Flora of Palestine" announced, *Ibis*, Series 5, l:130

**1884**
Exhibition and Remarks upon Some Specimens of Species of the Genus *Pachycephala* which appear to have been ignored or wrongly united to other species in a recently published
Catalogue of Birds of the British Museum, *Proc.Zool.Soc.Lond.* (Scientific Meeting), 2
On a Collection of San Domingo birds, *Ibis*, Series 5, 2:167–8
Notes on the Eighth Volume of the 'Catalogue of birds of the British Museum', *Ibis*, Series 5, 2:392–403
"Fauna and Flora of Palestine" noticed, *Ibis*, Series 5, 2:464
*The Survey of Western Palestine. The Fauna and Flora of Palestine*, Published by the Committee of the PEF. (Publication No.3), London.
*The Fauna and Flora of Palestine*, London.

**1885**
On Two Birds from Norfolk Island, *Ibis*, Series 5, 3:48–9
On a Small Collection of Birds from Korea, *Ibis*, Series 5, 3:194–5

**1886**
On an Apparently New Species of Duck (*Dafila modesta*) from the Sidney Isles, Central Pacific, *Proc.Zool.Soc.Lond.* (Scientific Meetings), 79–80
On the Species of the Genus *Plotus* and their distribution, *Ibis*, Series 5, 4:41–3

**1887**
On an Apparently New Species of *Zosterops* from Madagascar, *Ibis*, Series 5, 5:234–5

The Polar Origin of Life considered in its bearing on the distribution and migration of birds. Part 1, *Ibis*, Series 5, 5:236–42

On an Apparently New Species of *Zosterops* from the Island of Anjuan. *Ibis*, Series 5,5:369–71

## 1888

Letter: On the Breeding Plumage of *Podiceps occidentalis*, *Ibis*, Series 5, 6:148–9

The Polar Origin of Life Considered in its Bearing on the Distribution and Migration of Birds, Part 2, *Ibis*, Series 5, 6:204–16

Notes on a Small Collection of Birds from Newala, East Africa, *Ibis*, Series 5, 6:265–6

## 1889

Exhibition of an *Emberiza cioides* sent in by Mr. Chase from Birmingham, *Proc.Zool.Soc. Lond.* (Scientific Meetings), 6

Ornithological Notes on the Island of Gran Canaria, *Ibis*, Series 6, 1:13–32

Note on a Small Collection of Birds from Kikombo, Central Africa, *Ibis*, Series 6, 1:224

Some Stray Ornithological Notes, *Ibis*, Series 6, 1:227

Note on *Emberiza cioides*, Brandt, *Ibis*, Series 6, 1:293–4

On a Small Collection of Birds from Louisiade & d'Entrecasteaux Islands, *Ibis*, Series 6, 1:553

On the Peculiarities of the Avifauna of the Canary Islands, *British Association Report*, 616

On a New Species of Chaffinch *Fringilla palma* from the Canary Islands, *Ann.&Mag.Nat.Hist.* (6) 3:489

Catalogue of a Collection of Birds belonging to H. B. Tristram. Printed at the Advertiser Office, Durham.

**1890**

Remarks on a Visit to the Rock of Zalmo, *Proc.Zool.Soc.Lond.* (Scientific Meetings), 402

Notes on the Island of Palma in the Canary Group, *Ibis*, Series 6, 2:67

**1891**

Letter: On Birds Observed during a Voyage to Japan, *Ibis*, Series 6, 3:470

On an Apparently New Species of Pygmy Parrot of the Genus *Nasiterna*, *Ibis*, Series 6, 3:608

**1892**

Letter: On Dr. Koenig's Paper on the Ornithology of the Canaries, *Ibis*, Series 6, 4:181–2

On Two Small Collections of Birds from Bugotu and Florida, Two of the Smaller Solomon Islands, *Ibis*, Series 6, 4:293

Note on *Nestor norfolkensis*, Pelz., *Ibis*, Series 6, 4:557

**1893**

On the Bird indicated by the Greek *'Alcyon'*, *Ibis*, Series 6, 5:215

Differentiation, Migration and Mimicry, *Zoologist* 51:316–86. Presidential Address to the Biological Section of the British Association, Nottingham, September 1893

On an Undescribed Species of Snipe from the New Zealand Region, *Ibis*, Series 6, 5:446

On Tristram's Grackle in Captivity, *Ibis*, Series 6, 5:476

Presidential Address, Section D, Report of the British Association for the Advancement of Science. Nottingham, 784–97. Also *Zoologist* XVII:361–86 & *Nature* XLVIII:490

*Field Studies in Ornithology*. Annual Report of the Board of Regents of the Smithsonian Institution, Washington. Reprint of 1893 Presidential Address, Section D, British Association

**1894**

On Some Birds from Bugotu, Solomon Islands, and Santa Cruz, *Ibis*, Series 6, 6:28

Letter: On the Validity of the Species *Coracias weigalli*, *Ibis*, Series 6, 6:320

**1895**

Note on a Roller (*Coracias garrula*) in Northumberland, *Zoologist*, 53:433

Description of a New Species of Finch of the Genus *Crithagra* from South-East Africa, *Ibis*, Series 7 1:129–30

On the Use and Abuse of GenericTerms, *Ibis*, Series 7, 1:130 and 276

Letter: On Fragments of *Numida reichonowi* in his Possession, *Ibis*, Series 7, 1:165

Letter: On the Occurrence of *Pagodroma nivea* and *Chionis alba* on the Antarctic Continent, *Ibis*, Series 7, 1:165–6

Letter: Correcting the name of Sir John Ross, as mentioned in his previous letter, to Sir James C. Ross, *Ibis*, Series 7, 1:297

Further Notes from Birds at Bugotu, Solomon Islands, with a Description of a New Species, *Rambles in Japan, the land of the rising sun*, London

**1897**

Letter: On the Birds of the Levant, *Ibis*, Series 7, 3:481–4

**1898**

On a Small Collection of Birds made in Socotra by E. N. Bennet, *Ibis*, Series 7, 4:248

Preface to E. Carrington, *The Farmer and the Birds*, London.

**1900**

Letter: On Layard's Type Specimens, *Ibis*, Series 7, 6:693

**1902**

Notes on a Curious Accident to a Kingfisher, *Zoologist*, 60:68

Letter: On the Cahow of the Bermudas, *Ibis*, Series 8, 2:525

The "cahowe" of the Bermudas, *Ann.&Mag,Nat.Hist.* (7) 9:447–8

**1904**

Letter: On the Birds of Palestine, *Ibis*, Series 8, 4:164–6

**1905**

*The Victoria History of the Counties of England,* W. Page (Ed.). *A History of Durham 1: Birds*, 175–91

Fauna and Flora in L. A. Whibley, *Companion to Greek Studies*, Cambridge, 23–6

**1906**

Tawny Owls on the Banks: An Appeal, *Durham County Advertiser*, Feb 16, 1906

# Index

Aarhus 34
Accentor, Alpine 8
Accentor, Hedge *see*
 Dunnock
Acre 110
Acre, Plain of 109
Acts, Protection of Birds
 1954-67 251
Adaptation 89, 181
Adaptive Radiation 176
Addenda 223
Address, Presidential 37,
 90, 92, 93, 191
Agapane 177
Ain Bahrdad 63
Ain Beida 81
Ain Elibel 60
Ain Feshkhah 118, 133
Ain Mudawarah, Basin
 of 127
Ain Terebeh 119
Akurah 139
Aldabra 168, 171, 172
Aleppo 159
Algeria 57, 58, 76
Algerian Desert 87
Algiers 34, 52, 53, 56
Almond, Hely
 Hutchinson 35
Almond, Rev. George 34
Aln 2
Alnwick Castle 207
American Eagle 48
*American Ornithology* 13
Ammonites 134

Amzeh 149
Anitab 160
Anti-Darwinians 192
Antioch 161
Arabs, Jericho 150
Arad 134
Archaeology Department,
 Durham 38
Archaeology Museum,
 Durham 38
Arctic 40
Arctic Circle 216
Arnon, Ravine of 146
Assyria 204
Astronomy 35
Atet 154
*Athenaeum, The* 93
Atlas Mountains 63, 64,
 81, 155
Atoll 181
Audubon, J. 13, 229, 230, 249
Auk, Great 38, 43, 96, 210,
 212, 245, 253, 268
Auk, Little 34, 36
Aurora Borealis 145
Australia 186
Avibase 17, 18
Avis 44
Avocet 155

Beisan 131
Babbler, Arabian 115
Babbler, Bush 150
Babbler, Fulvous 62
Baker, R. A. 6, 37

Banias 134, 155
Barnaby-Lutley, Mr. 107, 130
Barrow-in-Furness 139
Bartlett, Edward 107, 118,
 125, 131, 134
Bartram, J. T. 15, 17
Bass Rock 4
Bateman, Mr. 130
Bath 3
Beagle 95
Beck, Rollo 231
Bedouin 131
Beebe, William 16
Bee-eater, European 56, 79,
 131, 159, 160, 161
Bee-eater, Persian 160, 161
Beersheba, 122, 123
Beirut 107, 113
Beni Sakkr 149
Benson, A. C. 97
Bermuda 8, 9, 13
Bermudez, Juan de 10
Berwick-on-Tweed 6
Beshni 161
Bethany 152
Bethel 113
Bethlehem 122, 126
Bible 95
*Bible, Natural History of
 the* 37
Bidwell, Mr. 214
Bill for the Protection of
 Wildfowl 252
Bird Architecture 191
Bird catcher 226

*Birds of Israel* 106
Birds of Paradise 187
*Birds of the World* 84
Birejik 160
Biskra 63
Bittern, Common 128, 135
Bittern, Little 128, 135
Black Watch 13
Blackbird, Red-winged 21
Blackbird, Siberian 207
Blackcap 150
Blad El Alma 81
Blaydon-on-Tyne 21
Blidah 53
Bluebird, Eastern 19
Bluetail, Siberian 207
Bodenheimer, J. F. 177, 224
Bodo 34, 246
Bolam, John 2, 217
Boscoe 246
Bou Hadjar 78, 79
Boucada 63
Boudisah, Sheik 58
Bou Guizoon 58
Boulac, 153
Bourbon 170
Bower Bird 186
Bowlby, Edward 28
Bowlby, Eleanor Mary 28, 40
Bowlby, Henry, 28
Bowman, H. T. 107, 111, 116, 120, 122
Brambling 34
Brandson, Jon 216
Brewster, Lady 95
Bridlington 254
Brighton 28
Bristol 3, 48
British Association 43, 47, 48, 93, 167, 191, 251, 254
British Association, Section D 47
*British Birds* 38
British Oological Association 250
British Ornithologists' Club 250
British Ornithologists' Union 46, 251, 267

Brush Turkey 187
Buckland, Frank 254
Buffalo 21
Bulbul 115, 172
Bulbul, Common 80
Buller, Sir Walter L. 185
Bullfinch 205
Bull Hotel 37
Bunting, Black-headed 155
Bunting, Cinereous 161
Bunting, Cirl 81
Bunting, Cretzschmar's 135, 155
Bunting, Ortolan 8, 155
Bunting, Snow 19
Bunting, Striolated 120
Bunting, Yellow 157
Burdon, Rowland 33
Burghuz 135
Bush Chat 60
Bussah, El 109
Bustard, Australian 186
Bustard, Great 214, 245
Bustard, Houbara 58, 59, 124, 132
Bustard, Little 82
Butler, Nathaniel, 12
Buxton, C. L. 141
Buzzard (Red-tailed) Long-legged 125
Buzzard, Common 2, 32
Buzzard, Honey 4
Buzzard, Long-legged 81
Byron, Lord 126

Cactus Finch 175, 231
Cahouze 11
Cahow 10, 12, 14, 15, 16
Caiffa 110, 111, 130
Cairo 153
Calle 77
Callirrhoe 150
Cambridge 46, 192, 213, 265
Cambridge University 39, 40, 41, 265, 266, 270
Cambridge University Dept. of Zoology 272
Canada 219
Canary Islands 166

Cannons, brass 147
Canon 37
Capercaillie 34
Carchemish 161
Carmel 125
Carmel, Mount 110, 130
Carolina 9
Carr, Ralph 2
Carre 170
Carribean 176
Carthage 76
Cassowary 188
Cassowary, Dwarf 188
Cassowary, Northern 188
Cassowary, Southern 188
Castle Eden 20, 28, 30, 36, 43, 46, 96, 214, 246
Castle Harbour 15, 16
Catalogue, Tristram's Bird 7, 13, 57, 78, 221, 226
Catalogue of Tristram's Books 266
Catbird, Grey 19
Cedar Scale Insect 19
Chamonix 8
Champley, R. 210
Chaplin, Henry, MP. 262
Chat, Arabian 113
Chat, Bush 60, 124
Chat, Red-rumped 60
Chat, Tristram's 60
Chat, White-headed Black 119
Chat, White-throated Robin 137
Chats, Desert 88
Che-Kiang 207
Chebka 52
Chevallier, Rev. Temple 37
Cheviot Hills 2
Chiffa 58
Chiffchaff 61, 107, 150
China 226
Chough, Yellow-billed (Alpine) 8, 137, 139
Church Missionary Society 202
Chusenji Lake 205
Close-time 255

# INDEX 313

Close-time Committee 256, 257, 263
Cochrane, J. H. 107, 130, 134, 136
Coco-de-Mer 166
Collectors 203
College, The 34, 37
Collins, J. C. 48
Comoro Islands 172
Company Colony 10
Coney 11
Conservation Movement 251
Constantine 80, 82
Constantino, Hotel 139
*Contributions to Ornithology* 13
Cooper's Island 14, 18
Coot, Common 55, 121
Coot, Crested 55
Coot, Flightless 170
Copenhagen University 217
Coquet Island 2, 4, 5, 30
Cormorant 207
Cormorant, Little 162
Cormorant, Pygmy 104
Corncrake 20, 108
Corsica 104
County Avifaunas 271
Courser, Cream-coloured 62
Cowpen Marsh 32, 33
Cragg, Prof. James B. 38, 211
Crake, Spotted 8
Crane, Common 124, 155, 206
Crane, Sacred 206
Crane, White-headed 206
Crane, White-naped 206
Crete 106
Cricket 36
Crossbill, Common 33, 60, 76
Crossbill, Red 21
Crossbill, Scottish 33
Crossbill, White-winged 21
Crossfell 31, 206
Crow, American 19
Crow, Carrion 31
Crow, Hooded 31, 112, 134
Crowley, P. 245, 248, 249

Crown Colony 10
*Cryptomeria* 205
Cuckoo, European 31, 76, 135, 149, 155, 183
Cuckoo, Great Spotted 76, 79, 130, 134, 135, 151, 155
Cuckoo-shrike 172
Cuckoo-shrike, Reunion 172
Cullingford, Joseph 38
Curator 37
Curlew, Common 31
Curlew, Eskimo 20
Curlew, Hudsonian 20
Curlew, Slender-billed 80
Cyprus 110

Damascus 138
Darrell, Richard 18
Darter 162
Darwin, Charles 46, 85, 89, 90, 92, 93, 95, 97, 98, 99, 104, 189, 193
Darwin, Emma 98
Darwinism 189, 190, 193
Data Cards 248
Deacon 8
Dead Sea 107, 113, 116, 118, 122, 133, 144, 155
Dead Sea, Map of 153
Decoy, Coatham Marsh 32
De la Warr, Earl 261
Denmark, 34
Derby, Lord 227, 273
*Descent of Man, The* 102
Desert Larks 9
Dhiban 147
*Dictionary of Birds* 212
Didine Birds 167
Dijon 7
Dipper 31
Disraeli, Benjamin 6
Diver, Black-throated 34
Diver, Great Northern 21
Djebel Dakma 79, 80, 81
Djelfa 60
Djem, El 76
Djendeli 80, 82
Dobson & Madeiros 19

Dodo 169, 172, 257
Dodo, Little 179
Dodo, White 170
Dodeersen 170
Dogs 172
Donors 223
Dotterel 4, 58, 124, 146, 147
Dotterel, Asiatic 110, 121
Dotterel, Red-breasted 110
Dove, Collared Turtle 117, 118, 145
Dove, Common Ground 19
Dove, Egyptian Collared Turtle 115, 134
Dove, Indian Turtle 145
Dove, Madagascar Turtle 171
Dove, Palm 113
Dove, Palm Turtle 113
Dove, Rock 19, 60, 81, 115, 121, 125, 128, 135, 146, 147
Dove, Stock 62
Dove, Turtle 63, 132, 134, 171
Draper, Prof. J. W. 93
Dresser, H. E. 255
Dreuse Emir 136
Dromedary 90
Drongo 172
Drongo, Papuan Mountain 180
Drummond-Hey, Col. H. M. 13, 40, 45, 46, 47
Du Bois 170
Duck, Bimaculated 148
Duck, Eider 30
Duck, Feruginous 57, 78
Duck, Labrador 226, 227, 228, 229
Duck, Marbled 135
Duck, Pied 229
Duck, Red-breasted Whistling 57
Duck, Skunk 229
Duck, Tufted 63
Duck, White-headed 57
Dundee 257
Dunlin 107, 118
Dunnock 31

Dunstanborough  4
Durham Cathedral  34, 37
Durham Cathedral Dean & Chapter  264
Durham City  3, 5, 28, 31, 136, 265
Durham City River Banks  264
Durham County C. C.  36
Durham School  5
Durham University  5, 6, 223, 224
Durham University Museum  210

Eagle, Bonelli's  77, 109, 126
Eagle, Booted  78, 81
Eagle, Golden  4, 56, 60, 61, 62, 80, 109, 114
Eagle, Imperial  77, 112, 121, 148
Eagle, Short-toed  80, 108
Eagle, Spotted  81, 149
Eagle, Tawny  79, 109
Eel  162
Egerton-Warburton, Mr.  107, 130, 131, 134
Egg Collecting  262, 264
Egg Registers  248, 249
Eglingham  2, 3, 5
Egret, Cattle  56, 57, 78
Egret, Great White  63, 128
Egret, Little  63, 128
Egypt  106, 54, 209
Eldey  166, 211, 214, 215, 218, 219
Eldeyjardrangr  215
Elias  131
Elliot, Sir W. H.  8
Elouibed, Well of  63
Embryology  191
Encyclopaedia Britannica  212
Endemism  186
Endor  112
Engedi  120
Ephraim, Mount  125
Eschricht, Etatsraad  217
Esdraelon  130

Esfia  111
Eshkol, Vale of  125
Euphrates  161
Evans, A. H.  218, 219
*Evening Star*  93
Exaltation  49
Exeter  8, 254
Exhibition, London  36
Eyton, T. C.  225

Faber, Frederick  6
Falcon, Barbary  61, 77, 79, 81
Falcon, Gyr  264
Falcon, Eleanora's  138
Falcon, Lanner  58, 121, 127, 128, 135, 144, 147, 151
Falcon, Saker  59, 63, 64, 148, 151
Falcon. Peregrine  2, 3, 32, 126, 204
Falconry  207
Fantail  180
Fantail, Desert  84
Fantail, Egyptian  107
Farne Islands  4, 30, 165, 216
*Fauna and Flora of Palestine*  106, 136
Fellowship, Drury  39
Fendi-y-Faiz  149
Feshkaha, El  118
Field Club, Tyneside Naturalists'  37, 90
Fieldfare  31, 114
Fiji  179, 180, 226
Fisher, James  230
Finch  174
Finch, Darwin's  176, 230, 232
Finch, Mangrove  232
Finch, Snow  137, 155
Finnmark  247
Fitzroy, Admiral Robert  95
Fjordland  186
Flamborough Head  4, 254, 257
Flamingo  170
Flamingo, Greater  58, 60, 76, 111

Fleming, Eleanor *see* Tristram, Eleanor
Flicker, Northern  21
Flower, Sir W. H.  191
Flycatcher  177
Flycatcher, Collared  156
Flycatcher, Narcissus  207
Flycatcher, Pied  31, 156
Flycatcher, Rarotonga Monarch  227, 230
Flycatcher, Spotted  31
Foochou Islands  202
Francis, Dr. William  43
Francolin  110, 158
Franklin Expedition  226
Froude, Mr.  6
Fuller, Errol  38, 212, 213, 215, 216, 218, 246
Fulling Mill  5, 38
Funk Island  216

Gadwall  78, 126
Galapagos Islands  174, 230, 232
Galilee  109, 126, 128, 129
Galileo  194
Gallinule, Allen's  58
Gallinule, Florida  18
Gallinule, Purple  55, 128, 135
Game Laws  254
Gannet  4
Garefowl  38
Garefowl Island  215
Geirfuglasker  215
Garganey  81
Garnet Head  15
Garnier  107
Gaskell, J.  217
Gaza  125
Gazelle  59
Geirfugle  214
Geistfuglasker  215
Geneva  7, 8
Genetic alteration  230
Gennesaret  126, 127, 130, 135
*Geographical Journal*  224
Gerar  125

# INDEX

Gerash 130
Gerazim, Mount 134
Ghardia 52
Ghor 121, 128, 129, 130, 132
Giacomo 129
Gibraltar 106
Gift Shop, Dilemma 211
Gifu 207
Gilead, Mount 134
Gisbourne, Rev. T. 210
Glaciations, Pleistocene 176
Gladstone, William E. 6
Godman, Alexander
  DuCane 40, 41
Godman, Percy Sandon 40,
  41, 225
Godwit, Bar-tailed 34, 246,
  247
Godwit, Black-tailed 126
Goldcrest 158
Goldfinch 113, 125
Good Words 166, 173
Goosander 170
Goose, Barnacle 135
Goose, Bean 154
Goose, Egyptian 108
Goose, Giant 182
Goose, Red-breasted 154,
  170
Goose, White-fronted 154
Goshawk 4
Govar, Bishop 36
Gowland, C. H. 244
Grackle, Tristram's 115,
  116, 119, 150
Grant, Ogilvie 250
Grant, Peter &
  Rosemary 176
Grantham 3
Grass-finch 186
Grassquit 176
Great Gun of Durham 268
*Great Sahara, The* 52, 64
Greatham 32, 36, 37
Greatham Hospital 32, 97
Grebe, Black-necked 55, 57
Grebe, Great-crested 55,
  121, 127, 128
Grebe, Little 32, 55, 112

Grebe, Slavonian 32
Greenfinch, Algerian 53, 80
Greenshank 8, 246
Grieve, Symington 212, 216
Gronvald, H. 268
Ground-finch, Large 231
Ground-finch,
  Medium 174, 231
Ground-finch, Small 231
Ground-nesting Birds 172
Grouse, Black 31
Grouse Epidemic 157
Grouse, Red 31
Guerah El Tharf 81
Guillemot, Common 30
Guillemot, Arctic Black 226
Gulf Stream 10
Gull, Audouin's 107, 109,
  118, 125
Gull, Black-headed 31, 55
Gull, Common 31
Gull, Great Black-
  headed 111, 127, 129, 130
Gull, Greater Black-
  backed 31, 34
Gull, Herring 31
Gull, Ivory 226
Gull, Lesser Black-
  backed 31, 106
Gull, Mediterranean 55, 107
Gull, Ross's 266
Gull, Royal Eagle *see* Gull,
  Great Black-headed
Gull, Yellow-legged
  Herring 107
Gunther, A. 38, 194
Gurney, J. H. 30, 40, 154, 225
Gurnot Rock 16

Habel, Dr. A 230
Hadj Khadour 131, 134
Haifa 107
Hakanarsson,
  Vilhjalmur 217
Hall, Fred. T. 16
Halloula (Huleh), Lake 53,
  54, 56, 135
Hammam Weled Zeid 77

Hancock, John 6, 37, 217,
  218
Hansen, Christen 217
Haram Wall 113, 114
Harford, Charles 7
Harrier, Hen 2, 3, 8
Harrier, Marsh 2, 108, 126
Harrier, Montagu's 57
Harrier, Pallid 130
Harris, Charles 231
Harris (Island) 246
Harrogate 28
Harting, J. E. 204, 251, 255
Harvard 213
Hawaiian Creeper 177
Hawaiian Islands 177, 178
Hawk, Hawaiian 227
Hasbeiya 135
Hayne, W. A. 144
Hazrun 139
Heath, Great
  Massingham 245
Hebron 122, 125, 132, 144
Hen, Mauritius Red 170
Henslow, Rev. J. T. 93, 94
Herbert, Sydney 6
Heredity 191
Hermon, Mount 137, 138
Heron 20
Heron, Black-crowned
  Night 56, 57
Heron, Buff-backed *see*
  Egret, Cattle
Heron, Green 18
Heron, Grey 8, 135
Heron, Purple 63, 135
Heron, Squacco 56, 78,
  126, 135
Hertford College 36
Hewitson, W. C. 217, 245,
  246
Heysham, T. C. 245
Highland Regiment 13
Hobby 138
Hogges, Wild 11
Holland, Harry 224
Holland, Rev. William
  Lyell 35
*Holy Land, Topography of* 37

Honeycreepers 178, 187
Honeysuckers 178
Hoopoe 8, 61, 77, 149
Hopegood, Private 16
House of Commons
  Committee 251
Honorary Doctorate 264
Hughes, Rev. Lewis 12
Huia 184
Hummingbird,
  Ruby-throated 19
Hurdis, Mr. 18
Huxley, T. H. 48, 94, 95
Hyrcanus 133
Hyksos 154

Ibis, Bald 160
Ibis, Glossy, 56
*Ibis, The* 37, 40, 43, 46, 48,
  85, 106, 108, 166, 224,
  251, 266, 267, 270
Ice Age 29
Iceland 5, 214, 217
I'Ive 177
Illness 6, 7
*Investigator* 226
Iraq 204
Ireland 219
Irruption 33
Island, Lord Howe 186
Island, Norfolk 186, 228
Island, Philip 186, 228
Islands, Aukland
  Group 186
Islands, Chatham
  Group 186
Islands, Cocos 232
Islands, Great Auk 165
Islands, Macquarie
  Group 186
Islands, Oceanic 20
Isle of Man 216
Iselfsson, Sigurdr 216
Isolation 168
Israel, the Land of 37
Italy 7, 219
Iyeyasu 203

"Jackals" 48

Jacamar 36
Jackdaw 5, 81, 114, 134, 159
*Jackson's Oxford Journal* 93
Jaffa 126, 144
Jamaica 226
Jamal, David 272
Jameel 132, 150
Jamestown 10
Jamrach's 38
Japan 201
Jardine, Sir William 6, 12, 13,
  18, 19, 20, 28, 179, 224
Jay 31, 161, 205
Jay, Algerian 53, 79
Jay, Black-headed 112, 130
Jebel Ajlun 130
Jebel Mushakka 109
Jebel Uslum 145
Jehalin 132
Jenin 112, 131
Jerba 112
Jericho 115, 116, 131, 132,
  152
Jersey 36
Jerusalem 36, 113, 114, 117,
  122, 126, 131, 152
Jesmond Dene 30
Johnson, R. C. 143
Johnstone, George 6
Jones, J. M. 14, 15, 18
Jordan 107, 117, 118, 129,
  132, 155
Jourdain National
  Collection 251
Jourdain Society 250
Jourdaine, Sylvane 10
Journal 40
Jubilee Meeting 46
Judah 125
Jungle, The 29
Juniper Scale Insect 19

Kadesh 125
Kaka 184, 227
Kammanga Fuchi 204
Kamschatka 204
Kanga, Mr. 226
Kedron 116
Kef 76

Kef Laks 76, 79, 80, 81
Keferein 133
Kerak 145, 147
Kestrel 30, 31, 62, 113, 172
Kestrel, Lesser 112, 114,
  128, 133, 151, 159
Ketilson, Ketil 217
Khan Miniyeh 128
Khifan M'sakta 81
Khorsabad 204
Kimmer Lough 2, 3, 4
"King Lion" 48
Kingfisher 186, 187
Kingfisher, Belted 20
Kingfisher, European 108,
  118
Kingfisher, Indian Blue 115
Kingfisher, Pied 108, 109,
  119
Kingfisher, Smyrna 128
Kingfisher, White-
  breasted 115, 119
Kingfisher,
  White-collared 233
Kingsley, Charles 169
Kirk, Dr. John 226
Kishion 111
Kiskadee, Great 19
Kite, Black 62, 78, 80, 132
Kite, Black-shouldered 58,
  109
Kite, Red 2, 4, 77, 81, 125
Kite, Royal 61
Kittiwake 30, 31
Kiwi 182, 185
Kiwi, Brown 185
Kiwi, Great Spotted 185
Kiwi, Little Spotted 185
Kiwi, Northern 182
Kiwi, Okarito 185
Kiwi, Southern 182
Kjaerbolling, Nils 212
Kjerringoy 34
Klein, Mr. 144
Korea 203
Kuweh, El 135

Lack, David 49, 214
Laghouat 61

# INDEX

Lamarck  96, 97
Lammergeyer  58, 77, 80, 128, 146
Land Crab  19
*Land of Moab*  118, 121
Langlands, J. C.  2
Lapwing  31, 120, 125
Larissa  159
Lark  86
Lark, Calandra  126, 147, 161
Lark, Crested  67, 84, 85, 87, 125, 126
Lark, Desert  62, 115, 123
Lark, Dupont's  62
Lark, Horned  138
Lark, Long-billed Crested  84
Lark, Persian Horned  137
Lark, Reboud's  84
Lark, Red-capped  158
Lark, Salvin's  84
Lark, Sand  63, 84
Lark, Shore  137, 138
Lark, Short-toed  85, 147, 155, 156, 157, 161
Lark, Temminck's Horned  61, 62
Lark, Thick-billed  89
Layard, E. L.  180, 226
Lebanon,  139, 158
Leeds  43
Legouat's Giant  170
Leontes  105, 155
Leontes, River  109
Leopard  54
Library, Tristram's  7
Library, British  12
Library, Palace Green  38
Lighthouse, David's  16
Lilford, Lord  225
Limestone, Magnesian  29
Lincoln College  6
Lincolnshire  36
Lindesfarne  2, 4
Linne, Carl von  84
Linnean Society  213, 214
Linnet  125, 137, 139
Lion  61, 68
Lisbon  106

Lisan  122
Listings, Tristram's  248
Liverpool  206, 222
Liverpool (Derbianum) Museum  223, 277
Liverpool Museum Catalogue  227
Living Landscapes  4
Local Names (Bridlington)  256
Lorikeet, Brush-tongued  187
Lorikeet, Duchess  233
Loretto School  36
Louisa Tristram *see* Tristram, Louisa
Lowne, B. T.  107, 111, 131, 134
Lubbock, Sir John  48
Lulea  247

MacGillivray, W.  246, 249
McGowan, R.  224
McLeod, Capt.  15
Madaba  151, 155
Madagascar  171, 172
Magdalene College  39, 40
Magpie, Common  31, 78, 159, 161
Magpie, Numidean  78
Malan, Rev. S. C.  250
Mallard  19, 32, 78, 148
Malta  106
Mamoud  139
*Manchester Guardian*  93
Manx Cat  85
Marquesa Islands  179
Marsaba  116, 117, 119, 133
Martin, Ash-coloured  121
Martin, Crag  58, 117, 119, 120, 121, 136
Martin, House  155
Martinez et al  77
Masada (Sebbeh)  120, 144
Mascarene Islands  167, 173
Mashita  148
Mauritius  166, 168, 171, 172, 226
Maxwell, Sir Herbert  263

May, Isle of  4
Meadowlark  21
Mearns, Barbara  5
Mearns, Richard  5
Mechlenberg  217
Medea,  58
Mediterranean  36, 52, 144, 222
Medjerdah  81
Medlycott, W. P. C.  106, 110, 126
Meidoum  154
Meiruba  139
Melika  62
Melville Islands  226
Merganser, Red-breasted  111
Merj-el-Ghuruk  126
Migration  191
Mikado, The  206
Mimicry  191
Missionaries  166
Moa, South Island  183
Moab  147, 152
Mocking Bird  174
Moho  178
Moller, Herr  217
Moorhen  18
Morchard Bishop  8
Moreau, R. E.  44, 46
Morocco  106
Morris, Francis O.  254
Mound Builders  187
Mountfort, Guy  43
Mowbray, Louis L.  16
Mowbray, Louis S.  16
Mullins, N. H.  244
Murphy, Grace  16
Murphy, R. C.  16
Museum, American, Natural History  16
Museum, British  228, 245, 248
Museum, British, Trustees  248
Museum, Glasgow  211
Museum, Hastings  249
Museum, Liverpool, Bulletin of  228

Museum, Liverpool World  30, 233
Museum, Norwich  80
Mussels, Fresh-water  135

Nablus  112, 126
Nablus, Pasha of  147, 148
Nadjar  161
Nagoya  207
Nahal Kesiv *see* Wadi Kurn
Nahr Seniku  108
Nahr-ez-Zaherany  108
Namibia  106
Napoloeon, Louis  48
Natural History Museum, Tring  251
*Natural History of Palestine*  106
*Natural History Review*  212
Natural Selection  88, 90, 97, 102
*Nature*  212
Navarone  106
Nazareth  112, 129, 130
Negev  123, 125
New Caledonia  180, 226
New England  12
Newfoundland  216
Newfoundland Bank  211
New Guinea  179
New Hebrides  179
New species  88
New World  21
New Zealand  179, 181
Newcastle-upon-Tyne  30
Newfoundland Banks  43
Newton, Edward  35, 40, 42, 167, 225, 226, 259
Newton, Prof. Alfred  14, 34, 35, 39, 40, 42, 43, 44, 46, 90, 91, 92, 93, 94, 95, 96, 97, 165, 166, 167, 188, 191, 193, 212, 214, 217, 218, 225, 251, 252, 253, 254, 255, 259, 265, 272, 273
Newton, Prof. Ian,  33
N'goussa  62
Niagara Falls  21

Nichols J. T.  16
Night Heron, Yellow-crowned  14, 19
Nightingale  53, 81
Nightingale, Palestine *see* Bulbul
Nightjar, Nubian  116
Nightjar, Red-necked  79
Nikko  204, 206
Nile  106
Noble, George  30
Nordland  246
Norfolk  245
North Africa  37, 83, 222
North America  20, 222
Northumberland  30
Northumberland, Duke of  255
Norway  34, 247
Norwood, Mr.  14
Notebook  248
Nutcracker  8
Nuthatch, European  136, 140, 158
Nuthatch, Great Rock  159
Nuthatch, Kruper's  136, 140
Nuthatch, Rock  158, 161
Nuthatch, Syrian  136, 140, 158

O'ahu Akepa  177
Oceanic Islands  162
Oede, El  67
Offoden Fjord  247
Old Bewick  2
*Ootheca Wolleyana*  212
Orde, Capt.  15
Oriel College  6
*Origin of Species, The*  85, 90, 96
Oriole, Golden  8, 135
Orkney Islands  216, 219
*Ornithological Biography*  13
Ornithological Congress  43
Osaka  201
Osprey,  106, 204
Ostrich  61, 62
O'u  177
Oudena  76

Ouzel, Black & White  207
Ouzel, Ring  138, 206
Owen, Prof. Richard  46, 94, 95, 214
Owl, Bare-legged Scops  233
Owl, Brown Fish  110
Owl, Eagle  8, 79, 114
Owl, Egyptian Eagle  112, 124, 133, 144, 148
Owl, Laughing  184
Owl, Little  62, 109, 113
Owl, Parrot  184
Owl, Pharaoh  112
Owl, Scops  8, 79, 114
Owl, Short-eared  3
Owl, Tawny  264
Oxford, Bishop of *see* Wilberforce, Bishop Samuel
Oxford Debate  93
Oxford University  6, 7, 28, 36, 210

Pacific Islands  223
Pacific Ocean  174
Palace Green  37, 38, 268
Palestine  36, 37, 106, 154, 222
Palestine Exploration Fund  152
Palestine, Map  141, 142
Palestine Papers  267
Palmer, R.  6
Papa Westray  216
Papyrus  210
Parakeet  172
Parakeet, Long-billed  228
Parasitism  191
Paris  7
Parkin, Thomas  249
Parliament  256, 267
Parliamentary Select Committee  258, 262
Parrots, Catalogue of  227
Partridge, Barbary  62
Partridge, Black  110
Partridge, Chukar  109, 118
Partridge, Greek  109, 125, 133, 145, 146, 151

Partridge, Grey  11, 31, 125
Partridge, Hey's  115, 133, 146, 151
Partridge, Red-legged  109
Partridge, Rock  144
Partridge, Sand  115, 118
Pass, St Bernard  8
Pastor, Rose-coloured  159, 160
Peed, Rev. James  53, 61, 75
Peel Island  135
Petra  116
Petrel, Bermuda  12, 16, 17
Petrel, Bulwer's  28
Petrel, Dusky  16
Petrel, Fulmar  5
Petrel, Leach's  227
Petrel, Mottled  16
Pheasant, Copper  203
Pheasant, Green  203
Pheasant, Ring-necked  203
Philadelphia Academy of Science  224
Philadelphia Museum  226
Philadelphia, Roman  133
Philippville  82
Pigeon  172
Pigeon, New Zealand  185
Pigeon, Rock *see* Dove, Rock
Pigeon, Tooth-billed,  179
Pigeon, Wood  111
Pigs  172, 174
Pimlicoe  11, 15, 16, 17
Pintail  32, 148
Pipit, Meadow  31
Pipit, Tawny  80, 81
Pipit, Tree  31
Plateau, Mascarene  168
Plover, Asiatic  124, 146
Plover, Caspian  124, 125
Plover, Golden  4, 20, 31, 110, 146
Plover, Green  10, 110
Plover, Kentish  58, 76, 110, 111, 126, 160
Plover, Norfolk *see* Stone Curlew
Plover, Ringed  126

Plover, Spur-winged  126, 128, 129
Pochard  32, 55, 57, 78, 118, 119, 121, 125
Pochard, Red-breasted  63
Pochard, White-eyed  63
Prairie Chicken  21
Praslin  166
Pratincole, Collared  58, 135
Pregez, Msr. Pictet de  7
Presentation copies (of books)  272
Prestwick Carr  2, 4
Procter, William  5, 6, 37, 210, 219
Protection of Seabirds, Association for  254, 255
Pryer, H. T. W.  205
Ptarmigan  34
Puffin  30
Purchas, Samuel  10, 14
Pylos  106

Quail, Common Bob-White  18, 19
Quail, New Zealand  185
Quail, Northern Bob-White  18
Queensland  188

Rabba  145, 146
Rabda  107
Racket-tail  187
Rail, Water  55
Rail, Weka  182
Rail, White-throated  171
Rasheiya  136, 138
Rat, New Zealand  184
Rat, Norway  184
Rats  172
Ratna  82
Raven  2, 3, 32, 59, 60, 114
Raven, Brown-necked  114, 118
Raven, Fan-tailed  120, 121, 143, 149, 151
Raven, Pied  87
Raven, Wedge-tailed *see* Raven, fan-tailed

Razorbill  245
Read, Mr.  210
Red Lions  46
Redshank, Common  31, 107, 110, 118, 121, 160
Redshank, Spotted  34
Redstart  8
Redstart, Black  109, 113, 119, 138, 139
Redstart, Moussier's  77
Redstart, Syrian *see* Redstart, Black
Redstart, Tethy's  109
Redwing  31
Reunion Islands  168, 171, 172
Reyjanes, Cape  215
Reykjavik  215
Rice, Nate.  224
Riflebird  180, 186
River Tees,  31
River Tyne  31
River Wear  31
Robin  107
Robin, Japanese  202, 205
Robin, White-throated  137, 138, 161
Rock Chat  85
Rock Thrush  8
Rock-Thrush, Blue  108, 113, 132
Rodriguez Island  168, 171, 172
Roller  79, 131, 133
Rook  114
Rothschild, Lord  232, 273
Rowley, G. D.  48, 229
Royal Navy  226
Royal Society  46, 193, 267
*Royal Society, Proceedings of the*  224
Royal Society for the Protection of Birds  251

Sacred Ibis  46, 209, 268
Sahara  87, 222
St Abbs Head  4, 5
St David's  16, 19
St George's (Bermuda)  15, 18

St Hild's College  38
St Kilda  5, 216, 219, 256
St Paul's School  36
Salmon, J. D.  6 , 213, 245
Salvadori, T.  188, 228
Salvin, Osbert  40, 41, 43, 47, 75, 76, 79, 179, 225
Samoa  179, 180
Samosait  161
Sandgrouse  59
Sandgrouse, African  138
Sandgrouse, Pallas's  37, 86
Sandgrouse, Spotted  117, 124
Sandpiper  118
Sandpiper, Common  31
Sandpiper, Green  63, 160, 246, 247
Sandwich !slands  177, 178
Sardinia  106
Sarepta  108
Sato, A.  176
Scandinavia  246
Scaup  31
Schultz, Fredrick  211
Sclater, Philip Lutley  40, 41, 44, 46, 48, 167, 177, 225
Scotland, National Museums of  224
Scrub-robin, Rufous  158
Sea, Dead  37
Sea Meawe  10
Sea Owle  11
Sea Venture  10
Seabirds Protection Bill  251
Seabirds Protection Act  258, 262
Seal Sands  32
Sealey, A. F.  41
Seaton Marsh  33
Sebkah  121
Sedgwick, Prof. Adam  95
Selami, Mr.  147
Selby, Prideaux John  6, 47
Select Committee on Wild Birds' Protection  259
Seran  188
Serin, Syrian  137
Serin, Tristram's  137, 139
Sewell, Mr.  217

Seychelles  166, 168, 171, 172
Sharpe, R. B.  2
Shearwater, Audubon's  15, 17
Shearwater, Dusky  15
Shearwater, Manx  110
Shearwater, Yelkouan  110
Shelduck  32, 121
Shepherd, C. W.  107, 116, 120, 121, 126, 127, 129
Sherburn Hospital  6
Shiloh  113
Shirahai, H.  106, 126
Shoveller  32, 78, 126
Shrewsbury  3
Shrike,  204
Shrike, Great Grey  109, 130
Shrike, Lesser Grey  155
Shrike, Masked  130, 155
Shrike, Pallid  61
Shrike, Red-backed  135
Shrike, Southern Grey  61
Sicily  106, 166
Sickle-bill  178
Sickle Bird  178
Silktail  180
Simpson, W. H.  41, 75, 76, 77
Skylark  8, 31, 125, 126, 147
Sloane, Sir Hans  12
Smith, John  12
Smith, Salton  15
Smithsonian Institute  16, 17, 244
Smyrna  136
Snipe, Common  20, 31, 63
Snipe, Great  258, 246
Society for the Protection of Birds  263
Society Islands  179
Solitaire  169, 172
Solomon Islands  179
Solomon's Pools  114, 125, 144, 149
Somers' Isles  10, 12
Somers, Sir George  10
Sotheby, Wilkinson & Hodge  270
Souf  63
Souk Harras  76, 77, 80

Sousa  76
Sparrow, Dead Sea  119
Sparrow, House  19, 29
Sparrow, Japanese Hedge  205
Sparrow, Moabite  150
Sparrow, Rock  138
Sparrow, Spanish  134
Sparrowhawk  204
Species  92
Starling, European  19, 31
Starling, Sardinian Black  126
Steendrup, Japetus  217
Stevens' Sale Room  247, 249
Stilt, Black-winged  58, 112, 146
Stint, Little  8, 63, 121
Stone Curlew, European  118
Stonechat  8, 172
Stork, Black  118, 161
Stork, White,  130, 132, 155
Strachey, William  10
Strickland, Mrs.  48, 49
Suf  129
Sunbird  172
Sunbird Hosea's  116
Sunbird, Palestine  115
Sunderland  31
Surgeons' Hall  214
Swallow, Barn  2, 61, 111, 155
Swallow, Bottle  203
Swallow, Oriental Chimney  111, 203
Swallow, Red-rumped  111, 117, 128, 131, 133, 203
Swallow, Rock  159
Swan, Whooper  114, 126
Sweden  247
Swift, Alpine  128, 134, 155
Swift, Common  137, 155
Swift, Galilean  115, 128, 129, 131
Swift, Little  117, 155
Swift, White-rumped see Swift, Galilean
Swinhoe, Robert  226
Switzerland  7, 221
Sylvian  86

Sykes, Christopher, MP 254, 255
Syria 102, 139, 161, 222

Tabor 131
Tanager, Summer 21
Tangier 106
Tatton 170
Taylor, E. C. 41
Taylor and Francis 43
Teal 31, 32, 78, 119, 144
Teal, Bimaculated 149
Tees Estuary 30
Tees Marshes 36
Teesmouth 32
Tegetmeier, N. B. 255
Tell Hhora 125
Temperley, G. W. 31, 32
Tern, Common 18, 19, 162
Tern, Gull-billed 107, 128
Tern, Least 19
Tern, Lesser Cresterd 55
Tern, Roseate 5, 18, 19
Tern, Whiskered 55, 57, 162
Tern, White-winged Black 58, 162
Thompson, William 254
Thornaby-on-Tees 31
Thrush, Blue Rock 77, 119, 135
Thrush, Hopping 119
Thrush, Mistle 31, 76
Thrush, Sociable Desert 62
Thrush, Song 31
Thrush, White's 31, 32, 33
Thrush-Nightingale 58
Tiberias 126, 127, 129, 130
Tibneh 130
Times, The 34
Tiritiri Matangi 186
Titmouse, Bearded 158
Titmouse, Blue 191
Titmouse, Coal 31, 139, 202, 204
Titmouse, Great 113, 202, 204
Titmouse, Long-tailed 31, 34, 204

Titmouse, Marsh 31, 202, 204
Titmouse, Sombre 139, 161
Titmouse, Varied 205
Tody 172
Tomes, R. F. 46
Tonga 179
Tornea 247
Torqay 34
Tortoise, Land 80
Tortuga Keys 249
Toth 209
Transmutation of Species 90
Tree-finch 175
Tree-finch, Large 231
Tree-finch, Small 231
Tristram, Charlotte 35
Tristram, Christiana 35
Tristram, Eleanor 30, 35, 47, 192
Tristram, Frances Anne 35
Tristram, Henry 35, 36
Tristram, Katherine 35, 36, 201
Tristram, Louisa 3, 6. 28. 30, 35. 36, 43
Tristram, Mary 35
Trollope, Anthony 169
Tropicbird, White-billed 17
Tropicbird, Yellow-billed 17
Trotter, M. 144, 148
Tuggert 63
Tunisia, Regency of 63, 64, 76, 106, 222
Tunnicliffe, C. F. 46
Turegs 90
Tyre 108

Um Rosas 147, 148
University of St Andrews 264
Upcher, H. M. 107, 116, 117, 118, 121, 125, 126, 129
Upolu 179
USA 10, 222

Varieties 86, 92
Vegetarian Finch 175, 231

Verreaux, Prof. M. 54, 57, 172, 206
Verrill, A. E. 12
Victoria College 38
Vireo, Blue-headed 21
Vireo, White-eyed 19
Virginia co. 10
*Virginea, History of* 12
Vivien 217
Vulture, Black 127
Vulture, Cinereous *see* Vulture, Black
Vulture, Egyptian 78, 79, 137, 147
Vulture, Griffon 76, 78, 109, 112, 115, 120, 127, 128, 129, 137, 146, 149

Waddi Sakk'r 151
Waders 20
Wadi Areyeh 120
Wadi Bireh 130, 131
Wadi Hamam 128
Wadi Heshban 133
Wadi Kelt 132
Wadi Kurn 109, 110
Wadi Mojab (Arnon) 146
Wadi Zuewirah 122
Wagtail, Grey 31
Wagtail, Grey-headed 80
Wagtail, Japanese Pied 204
Wagtail, Pied 31
Wagtail, White 113
Wagtail, Yellow 31
Wallace, Alfred Russell 35, 89, 187, 189, 193
Wallcreeper 8, 128, 136, 159
Warbler-finch 231, 232
Warbler, Aquatic 58
Warbler, Black-throated 144
Warbler, Cetti's 54, 55, 131, 138
Warbler, Cyprus 144
Warbler, Fantail 63
Warbler, Garden 31
Warbler, Graceful 116
Warbler, Greater Reed 58
Warbler, Grey-backed 158
Warbler, Icterine 56

Warbler, Melodious  54
Warbler, Olivaceous  54, 55, 56, 135
Warbler, Olive Tree  55
Warbler, Pallid  54
Warbler, Reed  55
Warbler, River  131
Warbler, Rufous  158
Warbler, Sardinian  54, 144
Warbler, Savi's  54, 55, 131
Warbler, Scrub  85, 120
Warbler, Sedge  31
Warbler, Sub-Alpine  53, 54
Warbler, Tristram's  63
Warbler, Upcher's  137
Warbler, Willow  31, 54, 107
Warblers  53
Ward, Rowland  249
Waregla  62, 76
Waterhen, Gigantic  181
Waxwing, Cedar  19, 21
Waxwing, Japanese  207
Wed Ballough  52
Wed N'ca  62
Wedderburn, Lt,  13, 20, 228
Wells, Thomas  250
Westerman Islands  214
Westfield College  202
Weymouth  3
Wheatear, Black  109
Wheatear, Black-eared  113
Wheatear, Buff-rumped  60
Wheatear, Desert  84, 118
Wheatear, Hooded  121
Wheatear, Isabelline  117, 161
Wheatear, Menetrie's  117
Wheatear, Mourning  89, 108
Wheatear, Northern  20, 60, 85, 103, 108, 114, 137, 158
Wheatear, Nubian  121
Wheatear, Red-rumped  84
Wheatear, Tristram's  63
Wheatear, Variable  108
Wheatear, White-crowned  119, 121
Whellan, W.  2
Whitaker, J. I. S.  266, 268
Whitburn Lizard  216

Whitby  28
White-eye  172, 183
Whitethroat, Lesser  139
Wigeon  32, 55, 63
Wilberforce, Bishop Samuel  93, 94, 95, 104
Wild Cats  12
Wildbird Protection Act 1972  261
Wildfowl Bill  258, 259
Wildlife Trust  4
Williams, J.  212
Williams, K. F.  250
Wilson, A.  13
Winchester College  36
Wingate, David. B.  17, 18, 19
Wise, J. de Capel  212
Witnesses (to Select Committee)  259, 260
Witthoos, P.  170
Wolf, European  124
Wolf, J. M.  44
Wollaston, A. F. R.  44, 49, 104, 213
Wolley, John  40, 43, 44, 46, 47, 96, 165, 214, 225, 243, 246, 247
Wood, Beanley  2
Wood Duck  21
Wood, Miss Ann  3
Wood Pigeon, Green  8
Woodcock  109
Woodlark  125, 126
Woodpecker, Black  34
Woodpecker Finch  174
Woodpecker, Great Spotted  79, 81
Woodpecker, Green  79, 81
Woodpecker, Lesser Spotted  79
Woodpecker, Middle Spotted  161
Woodpecker, Syrian  112
Woodpecker, Three-toed  34
World War, First,  38
Worrall, Emmeline  35
Wren  31

Yarrell, William  6, 46
Yokohama  226
York  3
York, Archbishop of  254

Zadam  149, 152
Zaharez  60
Zakinthos  106
Zanzibar  226
Zatun  149
Zib  109
Zimmer  129
Ziza  148
*Zoologist*  15, 34, 36
Zoological Society of London  255
*Zoological Society of London, Proceedings*  47, 166, 270
*Zoological Society of London, Transactions*  47, 255, 270
Zoological Society of New York  15, 34, 36
Zoology Dept., University of Durham  38
Zouave  54
Zouia  63